MAKING YOUR OWN TELESCOPE

Plate I. The author's 6-inch f/9 telescope, fashioned in accordance with the instructions in this book. The instrument weighs 52 pounds, and cost less than 30 dollars.

MAKING

YOUR OWN

TELESCOPE

BY

ALLYN J. THOMPSON

SKY PUBLISHING CORPORATION
Cambridge, Massachusetts

First printing, 1947

Revised, 1973

Eleventh printing, 1980

ISBN 0-933346-12-3

Printed in the United States of America

PREFACE

IN 1934, A SMALL GROUP OF YOUNG MEN, imbued with a mutual interest in astronomy and a curiosity about the tools of the astronomer, organized themselves into the New York Telescope Makers Association. With the erection of the Hayden Planetarium in 1935, a new focal point of astronomical interest was created for the New York metropolitan area. Shortly thereafter the telescope making group became a part of the Amateur Astronomers Association, an organization now of about 500 members, sponsored since 1927 by the American Museum of Natural History. Facilities and space for the telescope makers were provided in the basement of the Planetarium, where they began activities as the Optical Division of the Amateur Astronomers Association. As such I learned of the group and was privileged to become a member.

The Optical Division presently undertook classes in telescope mirror making, an activity which had been expanding from earlier beginnings at the Museum. With the advent of the war and the absence of some of our most skilled and active members, your author was asked to assist in carrying on with the teaching program, and has continued in this capacity to the present time.

In these classes, numerous problems arose that interest but perplex the tyro concerning the telescope and its parts, its functions and its optics, its capabilities and its limitations, and other related matters. The answers to many of these problems are as widely dispersed as amateur astronomers themselves, and are probably not to be found in any single volume in the consecutive and integrated arrangement which would be most useful to the beginner. Hundreds of mirrors in various stages of incompletion, made by amateurs working independently at home, have been brought to the author to have their ills analyzed.

In the light of these experiences, it seemed that the sort of service and instruction rendered at the Planetarium might worthily be made available to a greater number of amateurs. Accordingly, a series of articles on telescope making was prepared, and published in *Sky and Telescope*. These articles were received with much favorable mention, and it was therefore decided to expand and publish them in book form. And so this volume was born. In it the author has attempted to guide the novice past those pitfalls and snares into which the untutored worker is likely to stumble. Some new techniques are described, enabling excellent optical surfaces to be fashioned in a shorter time than heretofore generally required. The other parts of the telescope have not been neglected, and considerable study is devoted to the design of the various supporting parts. Descriptions of simple pipe mountings of proven efficiency are included. Endeavor has also been made to supply the answers to many questions not elsewhere treated. Some of the diagrams previously used have been redrawn and the number of illustrations has been more than doubled.

We are indebted to Earle B. Brown, an associate in the Optical Division, who read the original manuscript of the *Sky and Telescope* series, and made many corrections and valuable suggestions. Acknowledgment must also be made to Charles and Helen Federer for their patient help in the arrangement of the material and for the order of presentation.

An expression of gratefulness is also made to the Amateur Astronomers Association and its officers, and the members of the Optical Division, all of whom by their interest and active support have helped make this book possible.

ALLYN J. THOMPSON

New York, April 15, 1947

CONTENTS

Picture Credits

Plate I and Fig. 28. Peter A. Leavens

Plates II through V. Elwood Logan, American Museum of Natural History

Plate VI. Arthur Kolins

Figs. 73, 77, 78, and 79. Drawn by Arthur Kolins

Fig. 93. Jack Smollen

Fig. 94. Robert Adlington

MAKING YOUR OWN
TELESCOPE

Chapter I

STORY OF THE TELESCOPE

PRIOR TO THE TIME of the telescope, man's view of the celestial universe was woefully restricted when compared with what now can be enjoyed on any clear evening with ordinary binoculars. There were visible to him then only the naked-eye objects, the sun and the moon, five of the planets, and on a clear night stars down to about the 6th magnitude, some 2,000 in all.[1] A few hazy spots could also be seen, and there would be an occasional comet. Completely unknown were the outer planets, satellites of the planets, Saturn's rings, and infinite numbers of stars and galaxies.

Yet, working without optical aid, early observers managed to make some amazingly accurate charts of the visible stars, and amassed the observations from which the laws of planetary motion were deduced. The principal instrument used in establishing star and planet positions was the quadrant, a device having a graduated arc, and a pointer that pivoted about its center. With it Tycho Brahe (1546-1601), Danish astronomer, and one of the keenest of all observers, was able to record the positions of stars to within one minute of arc — about 1/30 the diameter of the moon. This

---

[1]There are about 6,000 stars in the sky that are bright enough to be visible to the average eye, but about half this number is contained in the celestial hemisphere that is below the horizon. Atmospheric haze obscures the fainter stars lying close to the horizon, thus reducing the number visible at any one time to about 2,000. Some authorities place the figure at 2,500.

1

was an amazing feat, when it is considered that one minute of arc is about the limit of visual acuity.

Then, in 1608, seven years after Tycho's death, the telescope was brought upon the scene by a Dutch spectacle maker, Jan Lippershey, to whom its invention is credited.[2] The invention marked one of the great progressive triumphs of man, enabling him to reach farther and ever farther out into space. It was not much of a telescope, this first refractor, consisting of two spectacle lenses perhaps an inch in diameter, one convex and the other concave, and magnifying possibly two or three times. Lippershey, whose name historians spell in various ways, managed to combine two such instruments into a unit, and thus also made the first binocular telescope.

The Galilean Telescope. Very soon, spectacle makers and scientists up and down Europe, learning of Lippershey's invention, were making similar instruments. Notable among the scientists was Galileo Galilei, the great Italian physicist and astronomer, who fitted a plano-convex and a plano-concave spectacle lens into opposite ends of a lead tube, making a telescope that magnified three times (Fig. 1). "They [the objects] appeared three times nearer and nine times larger in surface than to the naked eye," wrote Galileo. He experimented further and improved this erecting telescope as well as was possible with simple lenses, carrying the magnification up to 30 or more. This was about the limit of its usefulness, however, on account of the great reduction in the size of its field of view.

Fig. 1. Galileo's telescopes.

[2]The invention of spectacles, which in the course of time led to the telescope, was due perhaps to one Signor Salvino Armato, according to an inscription on his tomb: "Here lies Salvino Armato d'Armati of Florence, inventor of spectacles. May God forgive his sins. The year 1317."

The general arrangement of Galileo's telescope is shown in Fig. 2. Ordinarily, rays from a distant object *AB* would, after refraction through the objective lens *O*, meet to form an inverted image *ba* in the focal plane, but by interposing the concave eye lens

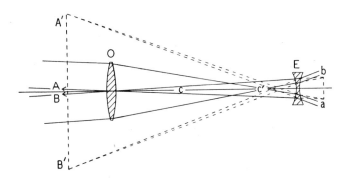

Fig. 2. Image formation in the Galilean (erecting) telescope.

E in front of that plane the rays are caused to become divergent, as though they had proceeded from the points *A'B'*, where a virtual image of the object is formed. This image is erect and magnified; the amount of magnification is the ratio of angle *c'* to angle *c*.

As the eye pupil can hardly embrace all of the rays emerging from lens *E*, only part of the actual field shown can be utilized. Also, the exit pupil (explained in Chapter IX) is located inside the instrument. The field of view thus depends on the size of the eye pupil, and on the diameter of the objective lens. The Galilean telescope is found today in the form of opera and field glasses, but employing quite moderate magnification: 2 to 3 power in the opera glass, and 3 to 6 power in the field glass.

The Keplerian Telescope. An improvement on Galileo's telescope was made in 1611 by Johannes Kepler, a German astronomer and former pupil of Tycho, who suggested that the converging rays from the objective be allowed to come to a focus, and that the resultant image be magnified with a convex lens. Fig. 3 shows the advantage of this new arrangement. The rays, upon emergence from the eye lens, are now converging; hence more of them and a wider field of view can be taken in by the eye. Projected backward through the eye lens, the rays appear to proceed from $B'A'$, where a virtual image, inverted and enlarged, is formed. As before, the

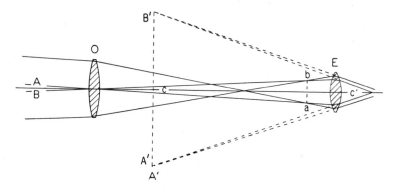

Fig. 3. Image formation in the Keplerian (inverting) telescope.

amount of magnification is in the ratio of angle c' to angle c. Considerably higher magnification can be had with this inverting telescope.

But with increasing magnification, the inherent defects of a lens, notably chromatic and spherical aberration (Figs. 4 and 5) were likewise increased. The aberrations could be diminished to a considerable extent by lengthening the focus of the objective lens. Consequently, in efforts to reduce these aberrations, enormous proportions were reached, instruments of 130 and 150 feet in length being constructed. Lens diameters up to six inches and more were attained. Non-spherical surfaces were also attempted in an endeavor to overcome spherical aberration. With these extremely long telescopes, working fields of only two or three minutes of arc must have been the rule. For comparison, the angular diameter of the planet Jupiter (at closest opposition) is almost one minute of arc, so the trials and patience of these 17th-century astronomers in aiming their exceedingly long instruments can be appreciated.

Magnification is a secondary consideration of the telescope; its chief function is to collect light. The eye alone gathers a limited amount of light, hence the luminosity of an object determines its visibility; also, the unaided eye can resolve only a limited amount of detail. An objective lens of the same diameter as the pupil of the eye would not improve vision, regardless of the amount of

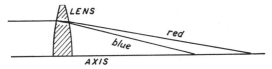

Fig. 4. Chromatic aberration in a convex lens. The amount of "bending" undergone by a ray of light depends on the refractive index of the glass. As the index is different for different wave lengths, and greatest for the shortest waves, a refracted ray of white light is dispersed into its component colors, blue being brought to a focus nearer to the lens than yellow or red light.

magnification employed, except that through this enlargement the detail in an object would be made more apparent. A 1-inch objective lens, assuming it to be about 3½ times the diameter of the eye pupil (about seven millimeters at night), collects about 13 times as much light, and correspondingly fainter objects become visible. The amount of detail seen is also increased, due to the greater aperture. So it is evident that the early observers needed larger objectives for greater light-gathering and resolving power.

But since spherical aberration increases with the square of the aperture, the only way in which it could be kept under control was to lengthen the focus, but there was a practical limit to what lengths could be handled. Moreover, a larger field of view was

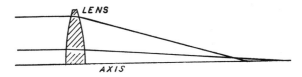

Fig. 5. Spherical aberration in a convex lens. Rays striking the different zones of a convex lens having spherical surfaces (or of a concave spherical mirror) are not brought to the same focus, edge-zone rays intersecting the axis at a point nearer to the lens (or mirror) than central-zone rays.

greatly desired, and this could accrue only with the use of shorter focal lengths. While spherical aberration could be pretty well eliminated by the use of two suitably curved lenses of the same kind of glass, there still remained chromatic aberration to be contended with.

In the hope of combining lenses of different glasses in such a way as to overcome chromatic aberration, Sir Isaac Newton attempted to determine if refraction and dispersion[3] were the same in all optical media. Although his experiment was inconclusive, from it Newton assumed that refraction and dispersion were proportional to each other, and he decided that nothing could be done to improve the refractor. He therefore directed his energies to the formation of images from concave reflecting surfaces, which are perfectly achromatic.

The Gregorian Telescope. Practical experiments with reflectors had already begun in 1639, but it was not until 1663 that they gained any prominence. In that year a Scottish mathematician, James Gregory, at the age of 24, published a treatise entitled *Optica Promota.* In this he gave a description of a compound reflecting telescope employing two concave specula (metal mirrors). The larger one was to be perforated, and to have a paraboloidal surface; the smaller was to be ellipsoidal.[4] The arrangement is shown in Fig. 6. Notice that the ellipsoidal mirror *s* is placed beyond the

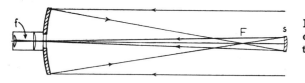

Fig. 6. Paths of light rays in the Gregorian telescope.

---

[3]Refraction is the bending of a ray of light as it passes obliquely from one medium to another of different density; dispersion is the change in the amount of refraction for light of various wave lengths.

[4]If an ellipse (see Fig. 34) is rotated about its major axis, the surface of revolution thus generated is a prolate spheroid, but telescope makers commonly use the general term *ellipsoid* for this figure. When an ellipse is rotated about its minor axis, the surface of revolution is an oblate spheroid. If a parabola or a hyperbola is rotated about its principal axis, the resulting surface of revolution is a paraboloid or a hyperboloid.

focal point F of the primary, which is also one of the foci of the
ellipsoid. From this position, the secondary mirror returns the
rays to form an erect image at its other focus f. High magnification
could be had with this instrument, the second reflection amplifying
the focal length of the primary in the ratio of fs to Fs. Construction
of the telescope was undertaken, but whatever chance it may have
had of performing creditably was lost by polishing the speculum on
a cloth lap — putty (tin oxide) being used as the polishing agent.
The unyielding lap was an insurmountable barrier to parabolizing,
interest apparently ebbed, and about 60 years were to elapse before
a workable model was finally produced.

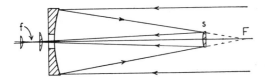

Fig. 7. Paths of light
rays in the Casse-
grainian telescope.

The Cassegrainian Telescope. Sieur Cassegrain, a French-
man, in 1672 designed a second compound reflector, differing from
Gregory's in that it employed a convex secondary, to be of hyperboloidal figure, placed inside of the focus of the paraboloidal
primary (Fig. 7). The image formed at f in this case is inverted.
The amplification from the second reflection is, as in the previous
case, in the ratio of fs to Fs, and while the Gregorian is seen to be
capable of higher magnification, all that is necessary can be had
from the Cassegrainian, and it has the advantage of being a much
more compact instrument. Although little was heard of this telescope for the next two centuries, it is worth observing that it sur-
vived the Gregorian, and is still widely used in observatories. The
principal reason for its early lack of popularity was no doubt due
to the difficulty of giving a hyperboloidal figure to the secondary
mirror. This difficulty can be avoided by leaving the secondary
with a spherical figure and undercorrecting the primary.[5] It is also
possible to leave the primary spherical, and to make all the correction at the center of the secondary, which will then have an oblate
spheroidal form.

---

[5]*Scientific American*, June, 1938.

The Newtonian Telescope. In 1668, Sir Isaac Newton designed and constructed a small reflector of the type so popular with amateur astronomers today and which still bears his name. His was not large, as we know telescopes today, the effective aperture of the concave speculum being about 1 1/3″. The focal length was

Fig. 8. Model of a 2-inch reflecting telescope made by Newton and presented by him to the Royal Society.

6″, making the focal ratio f/4.5.[6] To bring the light rays to a convenient place for observation, a plane speculum was used for the secondary reflection. This was placed at a 45° angle a short dis-

[6]The f/ number of a telescope mirror or lens is the ratio of its focal length to its diameter. In the above case, the focal length of the mirror is 4.5 times its diameter. A mirror of similar diameter but greater focal length would be said to have a higher focal ratio.

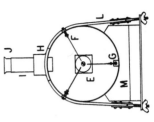

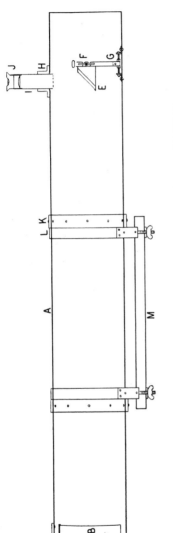

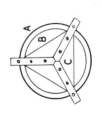

Fig. 9. SECTIONAL AND END VIEWS OF A MODERN NEWTONIAN REFLECTING TELESCOPE.

Above are the tube and component parts. Shown at left and right are suggested designs for the mirror cell and the spider support. A, tube; B, primary mirror; C, mirror cell; D, adjusting lock nuts; E, secondary mirror (diagonal); F, spider support; G, 45° adjustment; H, adapter tube holder; I, adapter tube; J, ocular; K, position rings; L, retaining straps; M, saddle.

tance inside the focus of the primary, where it could deflect the rays out through an opening in the side of the tube. There the inverted image was magnified with a plano-convex eyepiece of about 1/6″ focal length, giving a total magnification of about 36 diameters.

The plan of a modern Newtonian is shown in Fig. 9, and the reflection of rays in Fig. 47. As with the compound telescopes, the primary mirror must be a paraboloid, since a spherical surface cannot reflect parallel rays, such as those from a star, to a single

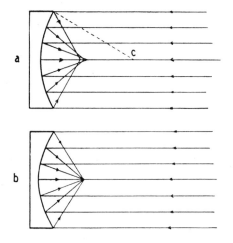

Fig. 10. Reflection of parallel light rays by: a, spherical mirror; b, paraboloidal mirror. The center of curvature of the spherical mirror is at C.

focus. This is shown in Fig. 10a, where rays striking the edge zones of a spherical mirror are brought to focus inside the focal point of the central-zone rays.[7] The paraboloid (Fig. 10b) is the only surface that can bring parallel rays to a single focus.

Newton, according to his *Opticks* (1704), polished his specula on pitch, using putty as the polishing agent. His methods were ingeniously calculated to yield a spherical surface, and it is quite probable that a close approach to that figure was attained. The performance of a spherical mirror of the small proportions of

[7] A zone is defined as a ring on the surface of the mirror, all points of which are equidistant from the center. It is often convenient, however, to regard a zone as having a finite width, notably the central zone which, on a 6-inch f/8, is substantially equal in width to half the radius of the mirror.

Newton's was probably satisfactory despite the small amount of spherical aberration present. Although Newton thought that his mirror might fail of good definition, he "despaired of doing the work" (parabolizing the speculum), yet he "doubted not but that the thing might in some measure be accomplished by mechanical devices."

Referring back to Fig. 10a, it might be concluded from a study of the diagram that if the center of the mirror were properly deepened, that is, given a shorter radius, or if the radii of the outer zones were progressively lengthened, or if a little of each were done, all the reflected rays could be brought to a common focus. That is a practical solution, and the resulting surface in each instance is a paraboloid. The standard practice is to deepen the spherical mirror so that, for a 6-inch f/8 mirror, the making of which is described in this book, the glass removed in the operation is but half a wave length of light in thickness at the center. Incredible though it seems, this represents the difference between poor and good definition.

The single-lens eyepiece of Kepler's had already been improved, with the addition of another element, by Christian Huygens, a Dutch astronomer and mathematician, about the year 1650. His compound eyepiece is shown in Fig. 68. The field lens, like Galileo's concave lens, is placed before the focal plane of the objective. As it is convex, however, it further converges the rays to form a slightly smaller image in a new focal plane, which is then magnified by the eye lens. Thus, a much wider field of view is encompassed by the eyepiece.

Further Developments. In 1722, John Hadley, an English mathematician, completed a Newtonian form of reflector (Fig. 11), in which the mirror evidently was suitably figured. It was about 5½″ in diameter, and 62⅝″ in focal length. With a mirror of these dimensions, practically perfect definition could be realized if the surface was given a spherical figure. (See the discussion on surface tolerance in Chapter VI.) This instrument attracted considerable attention, and presently other⁄makers were turning out Newtonian reflectors, following Hadley's technique, which consisted of removing the spherical aberration as it was revealed by the extra-focal diffraction rings of a star image.

Hadley then turned his attention to Gregory's design, and in 1726 he produced an instrument slightly over 2″ in diameter and 12″ in focal length. This proved so successful that construction was undertaken by other opticians, or artists, as instrument makers and craftsmen appear to have been then known. Notable among these was James Short, who made both Newtonians and Gregorians in great numbers, from about 1732 to the time of his death in 1768. Observatories purchased his larger instruments, a tribute to his skill, and the smaller ones were marketed chiefly among the aristocracy and those amateur astronomers of the day who could afford them.

The principal attraction of the Gregorian design was the erect image it gave, which made it suitable for terrestrial use. This cir-

Fig. 11. The first practical reflecting telescope, made in 1722 by John Hadley.

cumstance influenced its preference over the Newtonian, notwithstanding the fact that its images must have been pretty dull. Well into the 19th century, however, the Gregorian rode a wave of popularity that no type of telescope has known, until overwhelmed in comparatively recent years by the flood of amateurs who have flocked to Newton's design.

From the time of the invention of the telescope, and the startling discoveries of Jupiter's moons and the rings of Saturn, interest in astronomy had become something infectious. Each new discovery was accorded the widest publicity, stimulating a desire among those of learning to gain at first hand a glimpse of these celestial wonders. It was not practicable as yet for the average individual to make his own speculum, but many contrived to fit spectacle lenses into tubes, much as Galileo had done some 150 years earlier. Those whose means permitted bought telescopes, and envied was the gentleman who possessed one of three or four inches aperture, by an "exclusive" artist. But, judged by present-day standards, many of those reflectors were tiny. There is one (maker unknown) in the Fugger Collection at Augsburg, barely 1″ in diameter and 6″ in focal length, that was concealed in a walking stick! Eyepiece lenses of 1/6″ or less in focal length were quite common.

The metal used in those early mirrors was an alloy of copper and tin, the usual proportion about 75 to 25, which could be given a beautiful polish. But the metal was extremely hard to work, and a prodigious amount of labor was involved in grinding and polishing the curve. To facilitate the work, the comparatively thin disks were cast to the approximate curve, the backs also being curved to give uniform thickness and equalization of temperature effects. Grinding was done on convex iron tools of similar radius, using emery, and sometimes sand. Polishing was done on a pitch lap, with rouge. Manufacturers usually devised their own machines to do the work of grinding and polishing. Except where the utmost perfection was imperative, figuring seems to have consisted for the most part of a final brief variation of the stroke, in an unguided attempt to concentrate the polishing at the center. Critical testing, undoubtedly seldom indulged in on account of its laboriousness, could as yet only be performed on a star. In reflective ability, speculum was only about 60 per cent efficient, and the surface tarnished rapidly, effecting a further serious light loss. This meant frequent repolishing, and repolishing meant refiguring.

It is interesting to inquire into the prices that were asked for telescopes in that period, the latter half of the 18th century. Listed below are prices and sizes of a few of the Gregorians made by Short, selected from his catalogue. Newtonians in similar sizes were priced only slightly lower.

Diameter (inches)	Focal length (inches)	Magnification	Price (guineas) *
1.1	3	18	3
1.9	7	40	6
4.5	24	90-300	35
6.3	36	100-400	75
18	144	300-1,200	800

*An English gold coin, issued until 1813, equivalent to 21 shillings.

By the beginning of the 19th century, amateurs were able to procure specula for the primary and secondary mirrors, for both Newtonian and Gregorian designs. These could be had finished, ready for mounting, or as rough blanks to be ground and polished, not at the amateur's discretion, but to the curve of the iron tools furnished by the maker. Of course, the amateur had no means of correcting the figures of his mirrors, or even of knowing what they might be, the one reliable method of testing not being common knowledge.

Herschel's Contributions. Despite the attendant difficulties, a number of very large specula were made, some of the best by William Herschel. Born in Hanover, Germany, Herschel settled in England in 1757, where he became interested in astronomy and later (1776) turned his attention to telescopes. Working entirely by hand, at first as an amateur, he practiced and developed his technique on a great number of Newtonian telescopes, and learned how to figure the mirrors far better than had any of his predecessors. He performed the polishing in the conventional manner, with the mirror on top, and used a sweeping, circular stroke for parabolizing.

Later, Herschel applied himself to the design now referred to as the Herschelian type, which had been proposed by LeMaire, a French scientist, in 1728. In this design (Fig. 12), the mirror is tilted so that the image is thrown to one side of the open end of the tube, where it can be examined in comfort, with the observer's back to the object, and without the introduction of a second reflection. This latter feature was of tremendous importance in the days of speculum, when 40 per cent of the light was absorbed in undergoing a single reflection. Of less importance, but nonetheless gainful, was the elimination of the harmful diffraction effects from the

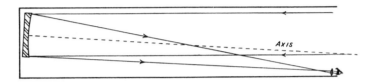

Fig. 12. The Herschelian reflector. Large mirrors might be tilted at but half the relative angle shown, as only a negligible fraction of the incoming light from a star would then be obstructed by the observer's head.

secondary mirror. But unless a suitably high focal ratio was chosen, astigmatic images resulted from the inclination of the mirror. And this introduced another problem; the lengthening of the tube meant placing the observer at an awkward height.

In 1789, Herschel completed his largest reflector, of the tilted-mirror type, which was installed at Slough, near Windsor. The speculum was four feet in diameter, with a focal length of 40 feet. It was about 3½″ thick, and weighed about 2,100 pounds. An elaborate and ingenious trestlework was built to carry the observer. (See Fig. 13.)

This great mirror was exceeded with the completion in 1845 of the largest of all specula, one six feet in diameter and 54 feet in focal length, by the Irish astronomer, Lord Rosse. The metal disk was nearly 6″ thick, and weighed about 8,380 pounds when cast. Rosse's gigantic instrument was mounted at Parsonstown, Ireland.

As representative of the prices Herschel charged for his reflectors, a Newtonian model of 6½-inch diameter and seven feet focal length sold for 100 guineas (30 guineas for the optical parts). Another 8.8-inch Newtonian, 10-foot focus, cost 200 to 300 guineas. Herschel advised buying two mirrors for this latter instrument (which probably accounts for the variable price) so that one could be used while the other was being repolished!

His talents were not confined to the making of fine specula; he also made his own eyepieces, some of which were truly remarkable. His frequent references to the use of magnifications of some 7,000 on his 6½-inch reflector occasioned some speculation and controversy among the English astronomers, but his claim appears to have been justified by the discovery, comparatively recently, of some very tiny eyepieces made by Herschel. Among his effects at Slough, W. H. Steavenson found several of these eyepieces, varying in focal length from about 1/16″ downward. The smallest of these was examined in a microfocometer, and found to have a focal length of 0.011″. It was bi-convex, about 1/45″ in diameter, and 1/90″ in thickness. It was tried out on a 6-inch refractor, and performed as creditably as its power would permit, but its field in that instrument was only about 20 seconds of arc in diameter. If Herschel actually used this eyepiece on his 85.2″ focal length reflector, it would have given a magnification of 7,668.

Fig. 13. Herschel's 48-inch reflecting telescope.

The Achromatic Refractor. In 1733, the achromatic lens (Fig. 14) was invented by Chester Moore Hall, an English barrister. This was accomplished by combining a convex crown and a concave flint lens in such a way that their focal lengths were inversely proportional to their dispersions. Although a number of telescopes were made according to Hall's instructions, the benefits of the achromatic lens do not appear to have been made available to the public until John Dollond invented it independently in 1758, and patented it. Dollond's efforts led to a demand for clearer glasses of more varied densities and of less equal dispersions, needed to improve achromatism, and chemists pursued experiments in learning

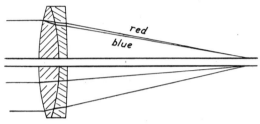

Fig. 14. The achromatic lens. The upper half of the diagram illustrates the color correction. The lower half shows how spherical aberration is corrected by the same compound lens.

how to control the refractive indices of melts, and in the pouring of large disks of limpid, homogeneous glass. Altogether, excellent progress began to be made, and by 1800 achromatic objectives 6″ in diameter were being turned out. Some of the best glass had been manufactured by Guinand, a Swiss who worked with Fraunhofer from 1805 to 1814. Fraunhofer produced a number of splendid achromats up to 9″ in diameter.

Dollond was making refractors (spyglasses) with single-lens objectives as early as 1742, his price for a 2-foot telescope then being 7s 6d.* In comparison, in 1762 he sold a 2-foot telescope with a two-lens objective (achromat) for 2 guineas. The lens diameters in each case were just under 2″.

In 1783, with a view to combining the benefits of the wide field of Huygens' eyepiece with a means of making micrometric measurements of an image in the focal plane, Jesse Ramsden, an English optician, designed the compound eyepiece shown in Fig. 68. It can be seen (more clearly, perhaps, from Fig. 65) that a

*In old English currency, 12 pence equaled 1 shilling; 20 shillings, 1 pound.

measuring device, such as adjustable parallel wires, set in the focal plane would be magnified along with the image. Measurement of an image in the focal plane was by no means a new idea; probably this had been first accomplished by Gascoigne, an Englishman, about 1638.

With the advent of the achromatic lens, the erecting or terrestrial eyepiece assumed considerable importance. This type of eyepiece was first suggested and used by Kepler, and improved in design about 1645 by Antonios Maria v. Schyrle, a Capuchin monk better known as Rheita. It is mentioned here because it spelled the rise of the refractor and the decline of the Gregorian for terrestrial use. As the terrestrial eyepiece has no significance in astronomical instruments, it will not be further discussed. Information on it can be found in any treatise on optical instruments.

In the early part of the 19th century, small achromatic refractors were being manufactured by several concerns. Prices, for apertures of 2″ to 3½″, ranged from 10£ to 50£. This included a pillar and claw stand, and in most cases, a rack work for focusing. With the larger sizes, three eyepieces of different powers were also included. A 5½-inch refractor by Dollond sold for 350 guineas. The lens alone could be purchased for 200 guineas. For those not having the means to buy achromats, telescopes with single-lens objectives continued to be made. Enterprising opticians were also offering lens sets that could be assembled into simple refractors.

The Modern Era. The method of chemically depositing silver on glass[8] discovered about 1840 by Justus von Liebig, of Nuremberg, was successfully applied to a small glass telescope mirror in 1856 by Karl Steinheil, a German physicist, and independently in the following year by Jean Foucault, the famous French physicist.

Then, in 1858, Foucault announced the development of his amazingly delicate and simple test for a concave reflecting surface, using an illuminated pinhole and a straightedge placed in the vicinity of the center of curvature of the mirror. The pinhole and straightedge were the outgrowth of earlier experiments in which simultaneous microscopic comparison was made of a pin point,

---

[8]Various processes of plating glass with metal for the making of mirrors had been known and practiced for centuries, but for one reason or another the coatings were unsuited for front-surface reflection.

likewise placed at the center of curvature of a mirror, and its reflected image, which was caused to fall alongside.

The last speculum of note to be constructed was one four feet in diameter, made by Grubb in 1870 for the Melbourne Observatory. Silver-on-glass mirrors replaced the more expensive and difficult-to-work speculum. Silver was far more efficient as a reflector, for a freshly deposited coat reflected somewhat better than 90 per cent of the light in the visible spectrum. It tarnished, but not nearly so quickly as speculum, and it could be removed by chemical means and a new coating applied without upsetting the figure of the glass surface.

Now, accompanied by Foucault's method of testing, reflecting telescope making by amateurs began to spread, slowly at first, throughout the civilized world. Books and articles on astronomy frequently contained instructions on the subject. Almost any thick disk of glass stood a fair chance of terminating its career at the bottom of a telescope tube. Porthole lights were especially preferred on account of their ready-made size.

The secret of the achromat was still the guarded possession of the professional optician, but simple lenses, in sets, and achromats as well continued to be offered on the market. Prices taken from catalogues of the year 1878 were, for single lenses, from two to six dollars for diameters of $1\frac{1}{2}''$ to $3''$. The higher-priced assortment included a "genuine Huygens eyepiece mounted in brass." Achromatic lenses, diameter $2''$, focal length $30''$, sold for $4.50, and 4-inch achromatic lenses of $60''$ focal length, with "extra fine finish," were priced at 75 dollars.

Almost the only obstacle now in the path of the reflector was the problem of casting and annealing larger glass blanks. These first glass mirrors, and even those of quite recent years, were of ordinary plate. The chief source of trouble was the high coefficient of expansion, with the resulting injurious effect on the figure caused by temperature changes. (Similar difficulties had been encountered with speculum metal, which had more than double the coefficient of expansion of plate glass, but the effects were no more pronounced due to the much thinner disks that were used. The thinner disks, however, were quite sensitive to flexure.)

Extensive experiment was undertaken in the United States, following the first World War, toward the development of a low-expansion glass. This resulted in a product called pyrex, first used for baking dishes. As is well known, pyrex was used in the casting of the 200-inch mirror blank, at Corning, N. Y., in 1934. Its coefficient of expansion is only one third that of plate glass, materially reducing deformation due to temperature changes. It is harder than plate glass and more difficult to grind, but it is also more resistant to scratches.

Following the successful pouring of the 200-inch disk, a clamor for this new glass went up from amateurs everywhere. The Corning Glass Company responded by providing suitably annealed disks in various sizes. It is fortunate that the homogeneity of the glass does not have to be taken into account. The melting point of pyrex is high, about 2,900° F., and even then it flows like cold molasses. As a consequence, folds, striae, bubbles and stones, in no way impairing its qualities as a foundation for a mirror, are to be found in almost any blank.

Fused quartz (silica), such as Corning's U.L.E. glass, has a much smaller coefficient of expansion and loses its heat faster than pyrex — distinct advantages in large professional instruments. Cer-Vit, Owens-Illinois' glass-ceramic material, has a "zero" coefficient of expansion and, like fused quartz, may be considered impervious to extreme temperature changes. But the high cost of these materials can hardly be justified in an amateur's first telescope.

At the close of the 19th century, experiments began on the deposition of metal films on glass by an evaporation process in high vacuum. These have resulted in the replacement of the silver mirror coating by a more durable one of aluminum. Although a number of metallic elements have been evaporated, some of them having a higher reflective index than aluminum for certain wave lengths, the efficiency of aluminum over the entire spectral range is higher than that of any other metal. It is but slightly less efficient in the visible spectrum than a fresh silver coat, with a reflective index of about 88 per cent, but it retains this reflective quality almost indefinitely. Its absorption in the ultraviolet is considerably less than that of silver, and so it is decidedly superior for photographic purposes. The first mirror to be coated by the evaporation process was the 15-inch of the Lowell Observatory. This was done in 1931 by Cornell University researchers, using chromium.

The largest refractor in the world today is the 40-inch telescope at Yerkes Observatory, Williams Bay, Wis. The lens, which has a focal length of 63.5 feet, was completed in 1895 by Alvan Clark. This is probably the ultimate in size for a refractor, because of flexures imposed by the weight of larger lenses, and also because of a serious light loss by absorption in the thick lenses.

The largest plate-glass mirror in use is the 100-inch, ratio f/5.1, at Mount Wilson Observatory. The glass blank was cast at the St. Gobain Glass Works in France and shipped to Pasadena, where it was made into a mirror by George Ritchey.

The 200-inch reflector, f/3.3, of the observatory on Mount Palomar, is by all standards the greatest of telescopes, a far cry from Newton's first reflector, 1 1/3" in diameter.

Reflector or Refractor? In the past there has been considerable discussion on the relative merits of reflector and refractor. From the standpoint of professional astronomers, there is no serious competition between them, as each type supplements the other in a well-rounded observing program. An amateur who plans to build his own instrument and to use it for general observing has other factors to take into consideration. Let us first look at some of the optical characteristics of reflectors and refractors.

Very early in the 19th century, when advocates of the speculum mirror began to feel the challenge of the refractor, Dr. Nevil Maskelyne, English Astronomer Royal, ventured the opinion "that the aperture of a common reflecting telescope, in order to show objects as bright as the achromat, must be to that of an achromatic telescope as 8 to 5." The relative inefficiency of the reflector of that day was due to the fact that, even under most favorable circumstances, barely 40 per cent of the original light escaped absorption by the metal mirrors, the greatest losses occurring in the short and medium wave lengths. Even silver-on-glass mirrors are subject to considerable deterioration, especially under certain conditions of the atmosphere.

The reflectivity of aluminum, however, is more-or-less constant, and from a standpoint of image brightness, it placed the reflector on a more equal footing with the refractor. In fact, until the development of anti-reflection coatings, an aluminized mirror had nearly the same efficiency, in light-saving characteristics, as an air-spaced achromatic objective lens[9] of equal aperture.

Coming down to figures — due to reflection there occurs in an untreated lens a light loss of slightly more than four per cent at each of its surfaces. With these reflection losses to be accounted for, plus an absorption loss in the substance of the glass (amounting to about two per cent for lenses of moderate size), it is evident that about 82 per cent of the original light is transmitted. In the reflector, after first deducting that area of the mirror's surface obscured by the diagonal, an equal percentage of the original light is found to be transmitted. Of course, this transmitted light is subject to another reflection by the diagonal, but the refractor will probably employ a star diagonal, the function of which is similar to that of the diagonal or prism of the Newtonian, so an equivalent loss may occur there. Therefore, with either instrument, the same amount of light reaches the eyepiece.

It was discovered, however, in the latter part of the last century, that some lenses which had been tarnished by the elements transmitted more light than ones that were newly polished; it was found that this resulted from lessened reflections at the tarnished surfaces. Various processes of producing an artificial tarnish were attempted. At present, in the most satisfactory method, metallic salts (such as magnesium fluoride) are evaporated in a high vacuum onto the glass. Ideally, the refractive index of an anti-reflection fluoride coating should vary from that of glass at the glass-fluoride surface to that of air at the fluoride-air surface, in which case no reflection would occur. Practically, the index of the coating should be equal to the square root of the index of the glass, and its thickness equal to a quarter of a wave length of yellow-green light. Only the light at opposite ends of the visible spectrum is then reflected, amounting in general to less than one per cent of that of the whole, and is detected by the purplish color given to the reflection.

The different fluorides vary in hardness, magnesium fluoride being perhaps the most durable anti-reflection coating for exposed

---

[9]It is common practice to cement together the components of small achromats when the adjacent convex and concave surfaces of the crown and flint lenses have the same radii (see Fig. 14), practically eliminating any reflection there. But on account of the unequal coefficients of expansion of crown and flint glass, lenses above 3″ in diameter are seldom cemented; thus an additional toll of light is taken at the two inner surfaces.

surfaces. The cost of application of the coatings is low, and thus improved, the refractor again moves up ahead of the mirror in light-transmitting qualities, a lens of about 5½-inch aperture being the equal, in this respect, of a 6-inch aluminized mirror.

In the matter of resolving power, the 6-inch mirror will excel the 5½-inch lens, although diffraction caused by the obstructive secondary mirror and supporting vanes of the reflector has a more-or-less deleterious effect on the image. The open ends of the reflector tube admit air currents which may further contribute to the impairment of image formation. Thus it may not be possible to realize the full theoretical limit of resolution of the mirror.

A considerable amount of correction of off-axis aberrations is effected in a suitably designed doublet lens; this cannot be accomplished in a mirror alone.[10] If the instrument is confined to visual use, however, these aberrations do not seriously handicap a telescope, as the observer naturally brings the object under observation to the center of the field, where definition should be limited only by diffraction.

On the other hand, the mirror is perfectly achromatic, while the doublet cannot be entirely freed of color. In order to reduce the residual color to a tolerable minimum, the ordinary refractor is usually designed in ratios from about f/15 to f/20, although, in the smaller diameters, it is sometimes made as low as f/10. The more versatile reflector may range from f/3 to about f/12, limited in the higher ratios by the accessibility of the eyepiece. In the short instruments, provision can be made for amplifying the ratio, when desired, permitting of observation at either the Newtonian or Cassegrainian focus.

Observing comfort with the 6-inch Newtonian demands that it be confined to a ratio of f/10 or less, in which sizes it can be made portable; in this respect the longer tube of a 5½-inch refractor may present a problem. Again, from the standpoint of comfort, the Newtonian has advantages over the refractor. The latter instrument, although equipped with a star diagonal, often places the observer in cramped positions.

Probability of success on his first attempt, cost, and upkeep are three important factors to be weighed by the amateur. On his

---

[10]See the discussion on *Aberrations of the Paraboloid* in Chapter XIII.

first venture, if he follows the instructions in this book with patience and care, he may expect to produce a reasonably good mirror; if it differs materially from the pre-ordained focal length, it does not matter. On the other hand, were he to attempt to make a refractor first, his lens would very possibly prove a total and expensive failure (good glass for a 5½-inch achromatic lens costs about 60 dollars). Any deviation from the prescribed radii of any of the four lens surfaces, by more than a very small per cent, increases seriously the amount of residual color.

As for cost, the excellent reflector shown in the frontispiece was constructed at a total outlay of only 30 dollars in the early 1940's, but would cost about three times as much today. A home-made refractor this size would run upwards of 400 dollars.

Upkeep is slight for any telescope, and least for a refractor. The reflective aluminum coating of the mirrors of a reflector is subject to deterioration from dust and the elements admitted by the open tube, but given the same protection when not in use that is accorded a refractor, at least two years of service should be realized before the aluminizing job need be repeated.

From the standpoint of an introduction to the optician's trade, the experience of thousands of amateurs has shown that one's teeth should first be cut on at least one *good mirror*. Then, if a refractor is contemplated, additional experience can be gained by making the optical flat that is so essential in the testing and figuring of the objective lens. (See Appendix B.)

What Size Reflector?   As already mentioned, telescopes are usually designed to perform particular kinds of work. Some are meant to be used chiefly for photography. In general, for visual work, low-ratio telescopes with their wide fields are useful for comet seeking, variable star work, and the like. The higher ratios are used in planetary study, double star observations, and in other fields where high powers and fine definition are required. Some of these instruments are portable, and others must be mounted on a solid pier. The amateur, however, usually will have formulated no particular plan of observation, except a desire to explore the heavens, and to see with his own eyes some of its wonders.

From the experience gained by amateur telescope makers, it has been found that the most practical and popular instrument for amateur use is the 6-inch f/8 Newtonian reflector. Its concave mirror is 6″ in diameter and its focal length 48″. The delicate task of parabolizing the mirror, while not easy, is not beyond the ability of a careful worker. The 4-foot focal length makes for comfortable observing, and with a low-power eyepiece, the field of view is a trifle over one degree in diameter — more than twice that of the full moon. The magnifications that may be employed permit of a modest size of mounting, which can be made portable. Such a telescope should reveal stars of magnitude 12.8, as compared with the 6th-magnitude limit of the unaided eye, and the 9th-magnitude limit of the average small binocular. Theoretically, the mirror is capable of resolving double stars having a separation of 3/4 of a second of arc, but as magnifications exceeding about 30 per inch of aperture can seldom be used, it may not be expected to perform up to this limit. This telescope will show the divisions in Saturn's rings; surface markings on the moon little more than a mile across should also be visible.

The purchase price of such an instrument of professional make is necessarily high, and many an amateur feels compelled to do without it. But if he is possessed of some ingenuity and craftsmanship, and is willing to devote a few hours a week to the task, he can in a relatively short time build the telescope in its entirety, for a small fraction of that price. Of course, many amateurs feel that their mirrors are inferior to the professionals', but this is not necessarily true. It has been frequently demonstrated that mirrors of professional make will seldom stand up to a test, because it is impossible for the professional optician to spend sufficient time on the mirror without losing money, whereas the amateur can, if he will, devote all the time and care necessary to produce a mirror of admirable figure.

In the following account, the amateur astronomer will be introduced to the optician's art of grinding and polishing spherical surfaces on glass. He will learn how to alter the concave spherical surface to a paraboloid, an achievement that baffled the world's opticians for 50 years, from Newton to Hadley. He will learn how to make and test an optically plane surface. He will find out why and how to make an equatorial mounting. He will take pride in his telescope, and derive profound pleasure from using it, because it will be a good one, and because he made it himself.

Plate II. Ages range from 16 to 70 in this mirror-making class. The man on the right is using a template to check a mirror curve; at left rear a student examines a mirror surface for pits.

Chapter II

MATERIALS AND EQUIPMENT

THE MATERIALS AND EQUIPMENT which are needed to produce the mirror are:

A *6-inch pyrex mirror blank,* from Corning Glass Co., Corning, N. Y. The sides of some pyrex blanks are tapered, so one face has a slightly larger diameter than the other. The larger surface is the one that is to be ground to curve for the mirror.

A *6-inch plate-glass disk,* of a thickness at least ⅛ the diameter, from any plate-glass manufacturer. This is to serve as the tool.

Abrasives (in the order in which they are to be used): carborundum, No. 80, 8 ounces; No. 120, 3 ounces; No. 220 and No. 400, 1 ounce each; alundum (fused aluminum oxide), No. 600 (or No. 2 garnet powder), ½ ounce; emery, No. 305 (or No. 8 garnet powder), ½ ounce.

Carborundum is a synthetic abrasive. The grains are separated into numbered sizes by passing them through a mesh of a corresponding number of strands per inch. It is considerably harder than emery, and in the coarse-grain size is about four or five times as efficient.

Emery, a variety of corundum (native aluminum oxide), is a mineral deposit. It is first broken up into powder form, and then graded by elutriation; that is, by repeatedly stirring in water and pouring or siphoning off the liquid, finer and finer settlings are obtained.

Garnet is also a mineral, and is treated in the same manner as emery. It is a cheaper product, but just as effective as an abrasive.

Polishing agents: choice of fine optical rouge (red), cerium oxide, or Barnesite, 4 ounces.

Red rouge (Fe_2O_3) is a product of iron oxide. This compound and to a lesser extent black rouge, another iron oxide, had for years been the only agents used in the polishing and figuring of fine optical surfaces.

Cerium, a metallic element, the oxide of which was discovered in 1803, was named for the asteroid Ceres, which had been found two years earlier. The oxide (CeO_2) is pale pink in color, has none of the messiness of rouge, and polishes two to three times as fast; many optical workers have recently used it in place of rouge. Some cerium oxide samples, on being sifted between the fingers, seem to contain rather coarse particles, but evidently in slow hand polishing, as on mirrors, these lumps disintegrate without doing any harm, as extensive trials have failed to produce a single scratch. In machine polishing, which is entirely different, cerium oxide does sometimes scratch, and the finer Barnesite is used for the finish polishing.

Barnesite (trade name), brown in color, is compounded of the oxides of several of the rare earths. It is finer than cerium, is perhaps as rapid in its action, and probably yields the most superior polish of all three products.

For a further discussion of polishing agents, refer to the section, *Theory of Polishing*, in Chapter V.

For the lap, about one pound of pitch should be obtained, as well as half a pound of rosin, and about two ounces of beeswax. Most optical pitch is a refined pine-tar product, a viscous fluid which, when cold, assumes a solid shape but flows under pressure. It usually has to be tempered with rosin or turpentine.

Rosin, beeswax, and sometimes a satisfactory grade of pitch can be obtained from a paint or hardware store. Pitch of doubtful purity, after being melted, should be strained through cheesecloth.

Sources of supply. Obviously, in purchasing grinding and polishing materials directly from manufacturers, large quantities would have to be ordered. It is much more economical to begin your first telescope with a mirror-making kit, available from many suppliers of amateur telescopes and accessories. In general, this is also a good way to avoid buying abrasives that have been accidentally mixed. Always dispose of any grade which should chance to become contaminated with a coarser one.

Mirror-making kits often include pitch that is properly tempered for immediate use without the addition of rosin or application of a beeswax coating. Some firms even supply a perforated rubber mat like that described in Chapter IV for making the pitch lap. Nevertheless, in keeping with the time-honored philosophy of amateur telescope making it has been judged desirable to retain in this volume some of the older methods, perhaps helpful to the isolated worker or the hobbyist who likes to start, insofar as possible, from raw materials.

Miscellaneous items: Also needed are a water pail; a water bottle or shaker, such as a cruet (or a tin can with a perforated cover); a coarse and a fine carborundum stone; turpentine; paraffin; a small paint brush about 1″ wide for applying the rouge; and a magnifying lens about 1″ to 1½″ in focal length. Also a small pot; empty cans for melting the pitch, beeswax, and paraffin; and an electric hot plate or gas burner.

To be made or improvised are: a template; a channeling tool; grinding stand or barrel; testing rack for the mirror; Foucault testing device; and stands to support the last two items. See page 36 for a list of mirror making materials and equipment.

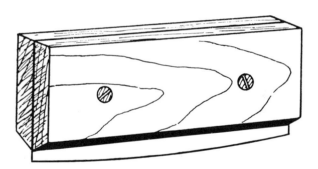

Fig. 15. The template.

The Template. This is used to gauge the curve on the mirror during coarse grinding. It should be given a radius of curvature of 98″, although the completed mirror should have a radius of 96″. The template is best made from thin sheet metal, 0.006″ to 0.010″ in thickness, and about 8″ square. To cut the curve, first nail the sheet metal to the floor. Drill two holes 98″ apart in the ends of a long stick, and nail one end to the floor so that the other end is free to swing across the metal sheet. Insert the point of a strong knife blade into the hole at the free end, and cut through the metal in a single sweep. With the magnifier, examine the convex edge for burrs, and carefully file them off. The template should be protected from bending by securing it between two thin boards, with its edge protruding about $1/4''$. Since a difference of 0.001″ in the sagitta of the mirror means a difference of about 1″ in focal length, great care must be exercised in testing with the template; watch carefully for any streaks of light that might show up between its edge and the surface of the mirror. See Fig. 15 for a diagram of the finished template.

The sagitta is the depth of the curve at the center, as measured between the mirror's surface and a straightedge laid across any diameter. It can be found with sufficient exactness from the value of $r^2/2R$, where r is the radius of the mirror, and R its radius of curvature. For our 6-inch f/8 mirror, it is 9/192 or 0.047″.

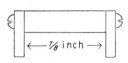

Fig. 16. The channeling tool in cross section.

Channeling Tool. This is used in forming the squares in the pitch lap (Fig. 27). It should be at least 8″ long, with a cross section like that shown in Fig. 16. A length of channel iron of these dimensions, or slightly smaller, will do, or the channeling tool may be made out of wood. It will not be needed if a molded lap is used. (See Chapter IV.)

Grinding Stand. Considerable friction is encountered in polishing, so the grinding stand must be rigid and heavy. A weighted barrel with a reinforced top will do nicely. A stand like that shown in Fig. 17 is sturdy, cheap, and easily put together. A 36″ to 40″ length of $2\frac{1}{2}''$ or 3″ pipe, screwed into a heavy flange, is set in a block of concrete. Another heavy flange is fastened to the heavy wooden top with small lag screws, and screwed tightly to the upper end of the pipe. A small cabinet in which to store the abrasives may be built in around the pipe.

Mirror Testing Rack. Fig. 18 shows a conventional form. The mirror is supported on two dowel sticks spaced about 4″ apart.

The rack rests on two pegs at the back, and on the adjusting screw in front, which is used to tilt the mirror up or down.

Testing Stands. Measurements will have to be made in hundredths of an inch, so the testing stands must be sturdy, and rest solidly on the floor. Any vibrations which result from passing trains or other traffic may sadly disrupt testing, so a concrete floor

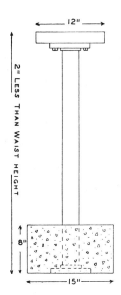

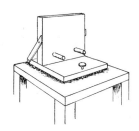

Fig. 17 (left). The grinding stand.

Fig. 18 (above). The testing rack for the mirror.

is preferred. If stands have to be built, they should be designed so that testing can be conducted in a comfortable sitting posture, and so that the center of the mirror, pinhole, and knife-edge are all at about the same level.

Foucault Testing Device. The requisites are a light source contained in an opaque shield in which there is a pinhole perforation, a knife-edge, and a scale for making measurements. Fig. 19 illustrates a practical testing device. The baseboard is a flat well-seasoned board, such as 3/4"-thick plywood, about 6" x 9" in size. The guide cleat, a straight-edged stick 1" wide and 6" long, is attached to the baseboard about 2½″ in from one edge. Fastened to this cleat in a suitable position is a strip of metal or plastic material on which has been scribed the scale shown in Fig. 20. The

Fig. 19. Foucault's test. In this device a prism is used to bring the pinhole and knife-edge into closer proximity. "Artist's license" is used in this depiction; actually, the illumination is too feeble to render the cones of light visible. The pinhole sends out light (via the prism) over a much wider angle, but only those rays within the defined cone fall upon the mirror; about four per cent of this light is returned by reflection to a focus at the knife-edge. For minute control of the latter, apply slight downward pressure at either of the spots, "X," as suggested in the text.

scale should be made under a reading glass for best accuracy. The horizontal lines are spaced 0.1″ apart, and the vertical lines 0.2″ apart. The diagonal lines divide the verticals into fiftieths of an inch, and interpolation can be made for readings to hundredths of an inch.

The knife-edge, of rectangular shape, and with a perfectly smooth edge, is cut from a thin piece of sheet metal, and tacked to a block of wood about 1½″ x 3″ x 3″ in size. If the lower part of the piece of sheet metal is cut and bent at right angles, as can be seen in Fig. 19, and attached to the block so that this edge just skims the surface of the scale in Fig. 20, it will also serve as the indicator. This indicator edge must be exactly parallel to the horizontal divisions of the scale when the block bears against the cleat.

For the light source, an ordinary 50-watt frosted lamp and socket are fastened to a 4″ x 4″ block of wood. A section of small stovepipe, or a tall can with both ends cut away, is set over the lamp. A "window" or hole about ⅛″ in diameter is drilled into the side of the can at a height opposite the bright spot of the lamp. A band of thin sheet metal, containing two or three pinholes of different sizes, is clamped around the can so that it can be slipped down over the window. Two pinholes, placed one above the other and about ¼″ apart, will suffice. To make these, thin an area of the metal band with a file and emery paper until it is only one or two

Fig. 20. The knife-edge scale, which should be made under a reading glass for greatest accuracy.

thousandths of an inch thick; lay it on a piece of glass, and, twirling a needle between the finger tips, pierce the first hole. Stop when the needle just breaks through, and examine the hole against a light with the magnifier to see that it is perfectly round. Any burrs should be removed with fine emery paper. For a finer hole, sharpen the needle on a fine stone, twirling and drawing it out at the same time under the finger tip, until it appears perfectly sharp under the magnifier. Pierce the hole as before, using less pressure. Holes about 0.005″ to 0.010″ in diameter are probably most satisfactory, but the size will depend on the amount of illumination used, which can, of course, be concentrated and intensified on the pinhole through the introduction of a condensing lens. Too small a pinhole may not admit sufficient light; a large hole is lacking in sensitivity. Foucault used a pinhole 1/12 of a millimeter (about 1/300 of an inch) in diameter.

To avoid discomfort from the heat of the lamp, and to bring the pinhole and knife-edge into closer proximity, a small right-angle prism may be used. Instead of the window of the lamp-container facing the mirror, turn it 90° to the left, so that it faces the knife-edge, and mount the prism as in the picture, on a block of wood, directly in front of the window or pinhole opening. A small chipped prism can be picked up for a few cents from dealers in optical goods.

A great many refinements are often added in an effort to increase the efficiency of this device, including an arc-light source, condensing lenses, extremely small pinholes or an adjustable slit, and knife-edge motions controlled in two directions by micrometer screws. But the seemingly crude device described should enable the reader to produce a mirror capable of the finest possible definition. Of course, the accuracy of the scale and one's ability to read it accurately determine its worth.

Nearly as simple to construct, and requiring no scale, is a small table, on which the knife-edge can be mounted, that rolls between guides on three ball bearings. A screw, which substitutes for the scale, controls the motion, with rubber bands or tension springs taking up the slack by an opposing pull. As the tension will have a tendency to lift the table from the ball bearings, the table should be weighted. Almost any thumbscrew will be suitable, or a piece of threaded rod into one end of which a pin is inserted at right angles, making a T-handle.

The number of turns and the number of threads per inch determine the extent of longitudinal motion. For example, consider a screw having 24 threads per inch, a standard number. This may be regarded as approximately 25 per inch, with a single revolution equal to about 0.04″, and a quarter revolution (easily estimated from the change in position of the T-handle), about 0.01″. Two and a quarter turns of a 24-thread screw will therefore yield an exact measurement of 0.094″, the full amount of correction called for on the 6-inch f/8 mirror (see Chapter VI for an explanation of correction), probably with a greater degree of accuracy than could be obtained from the first device.

Theory of the Foucault Test. This method of testing a concave reflecting surface was devised by Jean Foucault, of pendulum fame. The impressive features of this test are its simplicity and extreme sensitivity. Foucault's theory was that if an artificial star (illuminated pinhole) be placed at the center of curvature of a spherical mirror, then all of the rays from it that fall upon the surface of the mirror will be perpendicular thereto and, by the law of reflection, will be returned along the same paths, to form an image of the pinhole on the pinhole itself. Of course, the image could not be examined in this position, but if the pinhole were shifted to one side (we shift it to the right), the image would be shifted a corresponding distance to the other. Now if a knife-edge be cut into this image at right angles, all of the rays will be at once intercepted, and if the illuminated mirror were being viewed

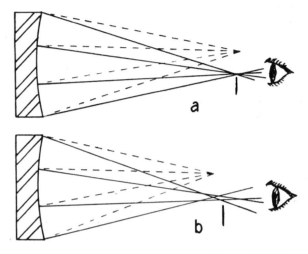

Fig. 21. Pinhole reflections from: (a) a spherical mirror; (b) an oblate spheroid.

by the eye as shown in Fig. 21a, it would appear to darken instantly. But if the curve on the mirror deviates the slightest amount from a true spherical figure, then all of the rays will not intersect in that pinhole image. Instead, rays from those zones of different curvature will intersect in slightly separated points along the axis of the reflected cone of light. One such variation of surface curvature is shown in Fig. 21b. Here, the knife-edge could nowhere be placed so as to intercept all of the rays at once, and to the eye viewing the mirror, those parts from which rays had been intercepted would appear dark, while the other parts would still appear illuminated. By exploring along a small range of the axis of the cone with the knife-edge, and measuring the distance between the points of intersection of these radii, it is possible to tell what our curve is, almost to the millionth part of an inch.

MIRROR MATERIALS CHECK LIST

Item	Quantity
1. **6-inch pyrex mirror blank**	1
2. **6-inch plate-glass disk**	1
3. **Abrasives:**	
Carborundum # 80	8 oz.
" #120	3 oz.
" #220	1 oz.
" #400	1 oz.
Alundum #600 or #2 garnet powder	½ oz.
Emery #305 or #8 garnet powder	½ oz.
4. **Polishing agent:**	
Rouge, cerium oxide, or Barnesite	
5. **For the lap:**	
Pitch	1 lb.
Rosin	½ lb.
Beeswax	2 oz.
6. **Miscellaneous items:**	
Water pail	1
Water bottle or shaker	1
Carborundum stone (coarse & fine)	1 ea.
Turpentine	
Paraffin	
Small paint brush	1
Magnifying lens (eyepiece)	1
Small pot	1
Empty cans	3
Hot plate or gas burner (kitchen stove will do)	1
7. **To make or improvise:**	
Template	1
Channeling tool (or rubber mold)	1
Grinding stand or barrel	1
Foucault testing device	1
Testing rack	1
Stands for testing device and rack	1 ea.

The various grades of abrasive listed opposite are by no means the only sizes that may be used. The transformation of the mirror's surface from a coarsely pitted one to the satin-like finish requisite for quick and complete polishing can be made with any series of progressively smaller grain sizes. It is expensive and time consuming to encumber the operation with many different grades, and yet if too large a gap exists between the various sizes, an undue length of time will have to be spent with each one. Hand grinding of a pyrex mirror is perhaps most efficiently accomplished with not fewer than six judiciously chosen grades. Use of a larger number is not necessarily a greater assurance of success, however, as there have been many instances where, with as many as 12 different sized abrasives being used, coarse pits still remained, and the last few stages had to be repeated.

It is not absolutely essential that the plate-glass tool be of the thickness recommended on page 27, but the thickness should never be less than 1/12 the diameter. For unless the surfaces of both tool and barrel top match quite perfectly, such a thin tool might yield slightly under the pressure of grinding, or the weight of cold pressing when polishing, and it would be difficult to maintain a surface of revolution. In any upside-down grinding, a thin tool is sure to bend under the pressure, and a spherical surface under these conditions is almost an impossibility. In considering the matter of thickness, see Chapter VII for further uses to which the tool may be put.

Plate III. Here is illustrated the proper position of the hands and arms, and of the body, while grinding. Pressure is the keynote for this operation.

Chapter III

MIRROR GRINDING

IT IS ASSUMED that a cellar or other room in which the temperature is fairly constant is available. Grinding, even polishing, might be done almost anywhere, but testing and figuring can be carried on only under conditions of uniform temperature.

Strokes. There are three motions that the optician must employ in order to preserve a surface of revolution on his mirror. First, the back-and-forth grinding stroke produces the curve. Second, the mirror must be rotated in order to produce this curve on all diameters. Third, the worker must walk around the barrel (or the tool must be rotated) in order to employ all diameters of the tool. While the directions of these last two motions are not important, they must not occur in unison, and for this reason it is perhaps safest if they are counter to each other. It is not necessary that any of the motions be performed with machine-like regularity, nor is it necessary to run oneself dizzy in their execution. Take six or eight strokes, rotate the mirror a trifle and shift to a new position where another six or eight strokes are taken, and so on.

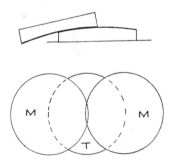

Fig. 22. The long stroke formerly used in rough grinding.

The lengths of the strokes at the disposal of the mirror maker are described with reference to the diameter of the mirror or tool. A *full-diameter* stroke is one in which the mirror travels a length of 6″; its center passes completely across the tool (Fig. 22). A *one-half* diameter stroke carries the mirror's edges halfway in to the

39

center of the tool, resulting in a travel of 3″. A *one-third* stroke
carries the mirror's edges 1″ past the edges of the tool. In using
these strokes, the center of the mirror need not necessarily pass
over diameters of the tool; much of the time, especially in the early
grinding, and finally, in polishing to figure, chordal strokes are
taken. But the length of the stroke — one-third or one-half, or some
other length — is constant, whether performed diametrically, or
over chords of the tool.

What Abrasive Does. If a grain of carborundum is placed
between two disks of glass and one disk is slid over the other,
a rolling action is imparted to the grain. As its sharp corners
impact the glass surfaces, fragments are chipped out, leaving pits.
The carbo grain is itself broken down in size until effective grinding
ceases. Increasing the rolling speed beyond a certain point merely
increases the rate at which the grain is broken down; in fact, the
chipping action on the glass is reduced. By increasing the pressure, however, the force of impact is increased, and larger chips
are broken out of the glass.

When grinding a telescope mirror, if the carbo is too thickly
applied, the separate grains are crowded together and are not free
to roll. Instead, they are dragged en masse around the work,
scratching and scoring as they go, until most of them have been
pushed over the sides of the tool, and the rest worn down to a size
where much of their efficiency is destroyed. It is a needless waste
to apply the abrasive thickly, and the amount of grinding obtained
will be small in proportion to the effort expended. Apply it thinly,
and use plenty of pressure, with emphasis on pressure. Speed of
stroke is also an important factor. Too rapid a stroke retards the
hollowing-out process, and will flatten the edge zones of the mirror.
Not more than 60 to 80 strokes per minute should be taken, calling
a stroke, in this instance, the combined back-and-forth motion.

Working in this manner, glass can be removed rapidly, and
with a minimum expenditure of elbow grease and carborundum.
Just enough water should be used to keep the work wet. It may be
necessary to add water from time to time as the abrasive breaks
down, or as evaporation takes place. In rough grinding to curve,
it is a waste of time to break each charge down fine. The purpose
here is to remove glass, and when the sound of the coarse grinding

ceases, a fresh charge should be added. The accumulation of "mud," consisting of crushed abrasive, glass, and water, should be flushed from the surfaces occasionally; otherwise it may cushion the action of the carbo and slow up the grinding. This is important when working with the finer abrasives.

How the Curve Is Obtained. In order to produce a concave curve on the mirror blank, it is ground face down on top of another glass disk of the same diameter, called the tool. By using diametric strokes of a length that will bring the edge of the mirror almost to the center of the tool (approximately the full-diameter stroke), then, at the end of the stroke (Fig. 22, top), the pressure per unit area of surface, and hence the abrasion, is greatly increased at the edge of the tool and the center of the mirror. Thus, a greater amount of grinding takes place in these regions, with the result that the upper disk becomes concave, and the lower one convex. It is interesting to note that gravity alone would bring about this condition without the application of pressure, but at a much slower rate. On account of the long stroke used (a shorter one would also work, but the action would be slower) the surfaces are not in contact, and are not spherical. The reason for this is that the edge zones of the tool comprise an area that is approximately three times greater than that of the center zones of the mirror, so there is a proportionately greater removal of glass from the center of the mirror. They "fit" together somewhat as shown in Fig. 23, which is exaggerated to show the hyperboloidal figure of the mirror, and the oblate spheroidal figure of the tool. After the curve has reached the proper depth, the surfaces are then brought into contact and made spherical by wearing back the edges of the mirror. This is accomplished by means of a short stroke, in the early use of which the radius lengthens slightly. The reason for this is apparent from a study of the diagram.

Fig. 23. The result of roughing out the curve with a long stroke: mirror hyperbolic, tool oblate.

It does not matter that the pyrex mirror blank and the plate-glass tool are of unequal hardness, and of different expansion coefficients. A more-or-less uniform "fit" of surface is maintained, depending on the stroke used, the plate glass merely being worn away at a more rapid rate. And for all practical purposes, the expansion difference is insignificant.

The foregoing is an old and reliable method of obtaining the curve, but requires that an unnecessary amount of glass be removed. The worker should grind his mirror by the quicker, more conventional method about to be described.

We Go to Work. The upper edges of the tool must first be beveled all around for about 1/16″, with the coarse carborundum stone. The stone can be applied vigorously to the tool, and if a few small chips are knocked off, which is unlikely, it does not matter. Later, as the surface is ground down, causing the edge to become sharp again, the beveling should be repeated. This is important, so that during grinding chips will not break off and get between the disks, causing scratches.

Have the pail of water placed conveniently near. Shake a few drops of water on the barrel top, and place on it several sheets of newspaper, with sprinkles of water applied between them, and lay the tool in the center. The wet paper will provide enough adhesion to prevent the tool from sliding about during grinding.

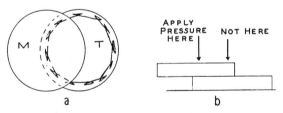

Fig. 24. The conventional method of rough grinding.

Apply some No. 80 carborundum around the edge zones of the tool, as that is where the grinding will begin. Add water, and place the mirror on the tool so that its center is about ½″ inside the edge. The one-third stroke is to be used, and grinding is to take place on chords of the tool, so that the center of the mirror is carried around the edge somewhat as in Fig. 24a. After taking six or eight chordal strokes (counting as one stroke the combined back-and-forth motion), rotate the mirror, shift to a new position, and work on another chord. The pressure should be applied over the center of the mirror (as shown in Fig. 24b), which should be sufficiently far in from the edge of the tool so that it will not teeter on the edge at the end of the stroke. If pressure were applied over the center of the tool, hollowing out would be slowed down considerably, although gravity would eventually do it for us.

After two or three trips around the barrel, wash and dry the mirror and try the template. When the curve in the small central area of the mirror seems right, take the next series of strokes on chords farther in on the tool. Spread the abrasive now over a wider zone, move the mirror in so that its center is about 1″ inside the edge of the tool, and grind as before. Apply plenty of pressure for quickest results.

Fig. 25. How the curve is obtained by use of the conventional method of grinding the mirror.

Frequent renewal of abrasive is necessary in the early stages, as much of it is pushed over the edge without being used. This residue can be scooped up later, placed in a jar of water and stirred, and the muddied water poured off. The unused abrasive at the bottom of the jar can then be reclaimed.

When the widened curve on the mirror fits the template, the mirror is moved inward another ½″ or so, and ground until the extending curve again conforms to the template. If at any time the curve becomes too deep, move the mirror in considerably, and continue with the one-third strokes. The curve should extend to within 1/16″ of the edges of the mirror when it is grinding center over center on the tool; the curve will have been achieved by the steps shown in Fig. 25. Of course, we can hardly expect the mirror to be spherical. The concave surface will probably consist of a series of waves, with its general curve almost inevitably slightly hyperbolic, somewhat like the condition shown in Fig. 23. This must be corrected by continuing with the one-third stroke, working center over center until contact has been established.

As the abrasive grains tend to collect in hollows and depressed zones, this stage of the work can be greatly facilitated if the disks are frequently separated and the carbo on both surfaces redistributed with the finger tips. At this stage, too, it is advantageous to prolong the time spent in grinding with each charge of No. 80 until it is thoroughly broken down. It may be observed, as the charges are ground down fine, that the mirror has a tendency to stick or grab whenever the grinding strokes bring the centers of the disks coincident. This sticking tendency occurs at the edge zones, and is the suction effect of the central gap between mirror and tool. Repeated breaking down to a fine consistency of each charge of No. 80, together with frequent respreading of the abrasive, and use of the one-third or shorter stroke, is the quickest way to grind the edges back and bring the surfaces spherical. When contact is established, the curve should extend practically across the whole diameter of the mirror, there being enough of the original bevel present to prevent a sharp edge. A more exact determination of the radius should now be made.

Flush off the disks in the pail of water, and wash your hands as well. Deposit about $\frac{1}{8}$ teaspoon of No. 400 carbo at the center of the tool, and puddle and spread it around the surface with the finger. Grind this down for about five minutes, stopping now and then to respread it. Then flush the mirror thoroughly and stand it, wet, on the testing rack. Using the window of the Foucault lamp, or any light source held alongside the eye, locate its reflection on the wet surface of the mirror. Move the light from side to side. If the reflection is seen to move in the same direction, you are inside the center of curvature of the mirror; if it moves in an opposite direction, you are beyond the center of curvature. When a position is found where the reflection just seems to appear, and from no determinate direction, as the light is moved from side to side, that is the center of curvature of the mirror's surface, and the distance from mirror to eye is to be measured exactly. One half of this distance is the focal length, which it should be possible to determine within $\frac{1}{2}''$ at this stage.

The charge of No. 400 enables the surface to hold the water for a longer period. Even then you may have to interrupt the test to splash more water on the mirror. If the focal length is found to be about 49'', you are ready for No. 120, but if it is too long, continue with No. 80 until you bring it right. It is preferable to use the one-third stroke for this shortening, but if the radius is too long by several inches, a longer stroke will speed up the process, after which some time must be spent in again restoring contact. If the focal length is less than 49″, invert the positions of the disks, and grind with the tool on top, using the one-third stroke. This stroke will lengthen the radius gradually, without deforming the figure. Use of a longer stroke, in any upside-down grinding, might flatten the mirror's edge to an extent that it might refuse to polish later on, unless contact were thoroughly restored by means of a short stroke, used with the mirror on top.

About two hours of actual effort will be expended in achieving a curve of the proper depth, with the surfaces brought into contact. While the difference of a few inches in focal length either way is of no material consequence, we have decided on a length of 48″, and the worker may as well begin now to develop a skill in working to precision, and endeavor to arrive exactly at that focal length for the finished mirror.

If through prolonged grinding the edge of the mirror has become sharp, it should be carefully beveled with the fine carbo stone. In doing this, rest the mirror on the edge of the water pail, and bevel at a 45° angle, using the stone as though it were a file. Rotate the mirror slowly, and continue the filing stroke until the edge is smooth and rounded to the touch. A coarse stone can be used on the tool; because of the wider angle of its edge there is little danger of chipping. Beveling will in all likelihood have to be repeated during the fine grinding, or whenever the edges become sharp and there is danger of scratches from chipping.

How to Determine Contact. Numerous bubbles collect between the disks during the grinding, with the larger ones at the center. As the abrasive is broken down, the surfaces approach each other more and more closely, and the bubbles become smaller. If those at the center remain large, or if they are concentrated locally in that area, they indicate a gap between the disks there. In this event, when a charge of abrasive has been broken down very fine, so that at the edges the surfaces may be actually abrading each other, suction caused by the central gap will make the disks stick and cling. Upon separating the disks, it will be found that the abrasive around the edges is fine to the touch, while comparatively large grains still remain at the center. When the rough-ground surfaces fit at every point, that is, when they are spherical, they will pass back and forth over each other evenly during grinding and no clinging will occur.

The means just described for determining contact are entirely satisfactory in the coarse-grinding stages, and in fact are all the criteria needed to carry one safely through fine grinding. But the mirror maker may feel more confident if he can have visible confirmation of what may be, in the case of the inexperienced, sheer guesswork. There are two ways of obtaining a visual check on the sphericity of the curve.

The first is a simple application of the axiom that only flat or spherical surfaces can be rubbed against each other and make contact at every point. Wash and dry both tool and mirror and, with the palm of the hand or under side of the forearm, brush off any lint that may be left on the surfaces by the towel. Now shortstroke the mirror on the tool, going through the motions of grinding for about half a minute. On separating the disks, a fine dust, consisting of powdered glass, will be seen in those areas where the surface has been abraded. Or if not clearly visible, the dust can be found and picked up on the tip of the finger.

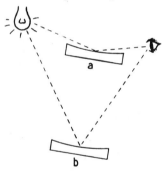

Fig. 26. Filament reflection test for sphericity.

The second method is a matter of variable reflection. We know that from a polished glass surface about four per cent of light of normal (perpendicular) incidence is reflected. As the surface is tilted, increasing the angle of incidence (the angle made with the normal), the percentage of light reflected increases, slowly at first, and amounts to total reflection as the angle of incidence approaches 90°. On an incompletely polished surface the amount of reflected light is lessened, and on a ground surface it may be nil (barring diffusion) except for very oblique rays, depending on the fineness of the grinding. And this is where use of a charge of No. 400 at the termination of work with No. 80, and the subsequent less-fine grades also, is beneficial.

Hold the clean, dry mirror face up, somewhat below a line between the eye and an unfrosted light bulb, as in Fig. 26a. A reflection of the filament will be seen on its surface. Now slowly lower the mirror, centering the reflection, which will become reddish or copper colored and fainter as the reflection angle is decreased. When it is quite dim, stop, and slowly slide the mirror to the right or left, allowing the reflection to pass across a radius, and off the edge. If the mirror is spherical, there should be just a slight and gradual diminution of the reflection from center to edge. In the event of imperfect contact (lack of sphericity), high zones will be revealed by a brighter reflection, and low zones by a dimming of the reflection. Slight differences can be tolerated, but any abrupt change in the quality of the reflection should be remedied before proceeding to a finer grade. In making this analysis while working on No. 80, the mirror may have to be held as at *a*, and in the final stage of fine grinding, as at *b*, Fig. 26.

Cleaning Up. Before going to fine grinding, or before going from any grade to a finer one, a thorough cleaning up should take place, so that there will be no possibility for a coarser grade to get near the mirror again. Otherwise, a grain or two of a coarser grade may find its way between the disks and chip wide swaths in the surface of the mirror. And, when polishing, even the finest grain of carbo on the lap will cause unsightly scratches.

The bench top, water bottle, spoon, template, flashlight, water pail, apron, and anything on which loose abrasive may lie should be dusted, wiped, or flushed. It is not necessary to remove every last carbo grain from the water pail each time it is flushed out. Well-behaved grains will remain at the bottom where they can do no harm. Flush the handle and rim, however, and keep clean water in the pail.

A skilled worker, abrasive-conscious and familiar with the routine, can bring a mirror through to the polishing stage without leaving his bench, save for checking the curve, pausing only long enough to change papers and to flush his hands and the disks. If you have not contaminated anything with carbo, no time need later be lost in an elaborate cleaning up.

Fine Grinding. The mirror maker has his choice of three ways of performing the fine grinding. The mirror may be ground on top of the tool, accompanied by a shortening of the radius; or the mirror may be placed on bottom with the tool on top, in which case the radius will lengthen; or the positions of the disks may be alternated, either with each charge of abrasive, or for half of the charges of each grade, and a constant radius can be maintained. By these means the worker has control of the focal length, which should be about 48¼″ when fine grinding is completed. Polishing will usually reduce this by ⅛″ to ¼″.

As to stroke length, the one-third stroke is the longest that can be used without deforming the surfaces. It will at the same time produce the greatest change in radius, amounting, if used continuously throughout fine grinding in the same positions, to 3″ or 4″ or more. A shorter stroke, say about ½″ in length, will produce a minimum of change, perhaps of only an inch or two. An advantage in alternating the disks is that greater freedom of movement is had, and equal abrasion is accorded to the edge of the mirror, where the least action takes place in the usual grinding-on-top. At no time should a rapid stroke be used in an effort to speed up the work, as this will almost certainly result in flattening the edge of the mirror. In any upside-down grinding, not more than a one-third stroke should be used, for a similar reason, except in a deliberate attempt to lengthen the radius radically. As he knows his focal length to a very close approximation, the worker can now select the stroke and method to be used with No. 120.

About six to 10 charges of each grade of carborundum, depending on the pressure used, should be thoroughly ground down, but for the beginner it is recommended that not less than eight charges be used. About five minutes is enough for each charge, but if one is unable to maintain heavy pressure a charge should be worked for a longer period. The purpose of each successively finer grade is to remove the pits left by the preceding grade, and to be assured of this not less than 40 minutes should be spent on each.

Before leaving No. 120, work the last charge down well, then follow with a charge of No. 400 (worked with the mirror on top), and again test for focal length. This time we have a surface which, when wet, will be capable of producing a fairly sharp image of the window of the Foucault lamp. Use a piece of ground glass, held before the eye, to pick up the reflection from the mirror, and adjust the lamp until it and the sharply focused image of the window are equally distant from the mirror. Half this distance is the exact focal length, which should be about 49″ at this stage. If it is found that a slight correction must be made, of an inch or two, it should be done on this grade. Any greater correction should be made with No. 80. It may be more convenient to use a piece of bright tin, rather than the ground glass, in picking up the reflection, as the worker may then stand closer to the mirror and can more quickly reach and adjust the rack. The eye has to be placed in back of the ground glass, but a more critical determination of sharpest focus can be made with it.

When working with the mirror on top and using a short stroke, there may occur a slight lengthening, rather than the expected shortening of the radius. It would have been difficult and impractical to bring the surfaces into exact contact with No. 80 on account of the coarse grain size, and a thin gap may have persisted at the center. But as finer grades are used and broken down, the surfaces approach each other more closely, and a slight lengthening of the radius may thus take place through the wearing back of the edge zones of the mirror. After complete contact is established, however, the radius will begin to shorten.

The focal length now being quite exactly known, the worker can proceed with confidence. The radius can again be checked upon the completion of each grade. As finer abrasives are used, the quantity for a single charge should be reduced; for example, just a pinch of emery No. 305 will suffice. In using carbo No. 400 and finer grades, the abrasive should be deposited at the center of the tool (or mirror, if it is on bottom), a few drops of water added, and the mixture puddled and spread over the surface with the finger tip. After each of these finer charges is thoroughly worked down, the surfaces should be flushed clean of mud before applying the next charge. Faster action is obtained in this way. With the last two grades, the tiny bubbles that have always been present between the surfaces should be eliminated before grinding starts. This is done by moving the mirror out over the tool, slowly rotating it the while; the bubbles will disappear over the edge, leaving nothing but a thin film of water and abrasive between the disks. Only five charges of No. 305 need be used, but the last one should be worked out for about 15 minutes, with the mirror on top, adding water when necessary. It should be possible to read 12-point type, held at a distance of several inches, through a mirror ready for polishing.

If the tool is of the recommended thickness, pressure can be used throughout the fine grinding, but beware of the disks sticking together in the final stages. Evaporation may cause them to stick, so if it is necessary to pause for a moment, the disks should be separated. Otherwise, keep them moving at all times, and if there is a suggestion of stickiness, due to evaporation, separate them immediately. As previously explained, suction from lack of contact at the center will also cause them to stick. If this occurs, use a carpenter's wooden clamps to separate them, or wedge the tool on a bench, hold a block of wood against the side of the mirror, and free it with a sharp blow of a hammer.

Examining for Pits. Hold the clean, dry mirror before a light, and with the magnifier examine the texture of the surface at the center. The pitted surface should appear uniform with no intermingling of large pits. Compare it with an area out near the edge. There should be the same uniformity of surface appearance, indicating that grinding is going on evenly. It may be helpful to cut up four small squares of plate glass and grind them together in this fashion: 1 and 2 with No. 400; 2 and 3 with No. 220; and 3 and 4 with No. 120. The worker is thus provided with comparison surfaces of these grades, but the same texture should not be looked for, as it should be remembered that there is a difference in the hardness of pyrex and plate glass. Prevention here is the best cure. If the worker thoroughly grinds down eight charges of each grade of abrasive, with pressure continuously applied, he need waste no time looking for pits.

Chapter IV

THE PITCH LAP

Making the Lap. We are now ready for the polishing lap. First, cut a strip of newspaper at least 20″ long, and 1/4″ wider than the thickness of the tool; run it through some melted paraffin and set it aside. This is to serve as a collar, to be wrapped around the tool to retain the hot pitch while it sets. Now melt the pitch slowly on the stove, in a pot or can. It is highly inflammable, so keep the flame low, and have on hand a board or other cover that can be immediately placed over the pot to smother any flames. Use a wide, thin stick for stirring. If a can is being used, it may be gripped with a pair of pliers while pouring. When the pitch is fully melted, pour a large drop onto a piece of scrap glass and let it cool. Try denting it with your thumbnail. If after about 15 seconds of pressure, there is hardly any indentation, the pitch is too hard; too soft if it has yielded easily. Some mirror makers use the bite test, although this is less determinate than the thumbnail test. Chip the cold drop of pitch free from the glass and place it between the teeth. If it immediately shatters into fragments under pressure, it may be too hard. There should be just the slightest yielding before breaking. The surest way to find out about the temper is to go ahead and make the lap, and see how it works in polishing.

To soften pitch, add just a few drops of turpentine. *Do not do this over the stove.* It takes very little turpentine for this, so use it cautiously. Stir thoroughly and again try the temper. To harden pitch, cook it for a long period. Adding a quantity of rosin will harden it. No two pitches are alike, so exact proportions cannot be given, but the amount of rosin added ought not to exceed the volume of pitch. Some laps are made entirely of rosin; these are most suitable for hot climates.

While the pitch has been melting, partially fill the pail (which has been thoroughly cleaned out) with water as hot as the hands

51

can tolerate, and completely immerse mirror and tool, standing them on their edges and resting them against the sides of the pail, so that they can be quickly reached when wanted. Have a piece of soap on hand in a clean dish. The tool and mirror should be al-

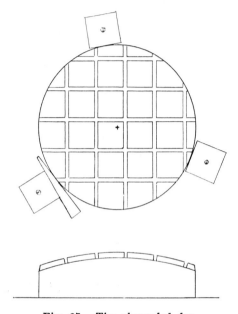

Fig. 27.   The channeled lap.

lowed to heat through before you pour the lap. After the pitch has been fully melted and tempered, remove it from the stove and allow it to cool slightly, as it is best not to pour boiling hot pitch directly onto the tool. Now remove the tool from the hot water, dry it, and wrap the paraffined paper collar around it. There should be enough heat in the glass to cause the overlapping ends of the collar to adhere. Give the pitch a final thorough stirring and pour it onto the tool up to the level of the collar. When it has thickened enough to retain its shape, remove the collar.

Soap the channeling tool and press the channels into the warm pitch right down to the glass, as in Fig. 27. Start at the center, and locate the first channels so that the center of the lap will lie just inside one corner of a square. This is to avoid a uniform radial distribution of the squares, and so prevent zonal rings on the mirror. Press the channels in one direction first, then press the other set at right angles. Take the mirror from the hot water, quickly soap its surface, and slide it about on the lap, gradually increasing the pressure. The channels will be closing in now, so replace the mirror in the hot water, and go over the channels again. Follow with the soaped mirror. These steps may have to be repeated two or three times before the lap has hardened sufficiently to retain its shape. By proper manipulation, any tendency for concavities to develop in the squares can be ironed out. By the time the pitch has hardened, the lap should exactly conform to the curve of the mirror. Remove the mirror and wash it off in clear water.

Using a sharp knife, with a firm shearing stroke cut away the pitch overhanging the edge of the tool. Incline the knife so as to give a slight bevel to the lap. The knife, or a razor blade, can be used to widen any channels that may have closed in. Do this cautiously or large fragments of pitch may break out. A good width for the channels is ⅛″. Flush the lap under a cold-water tap to clean off the soap and any pitch fragments, and dry it. Melt down some beeswax in a can. Prepare a brush for the wax: a thin stick of wood the same width as the squares, with a chisel-shaped edge, over which three or four thicknesses of gauze bandage or similar material are wrapped and tied with string. Soak this in the melted beeswax (which should be smoking hot) and draw it across the lap, covering each square with a single stroke of the brush. Two or three squares can be done on a single dipping. While the mirror might be polished directly on the pitch or rosin lap, the beeswax coating gives a faster, sleek-free polish, and results in a better edge. (Sleeks, or micro-scratches, faintly visible on a polished surface, may result from coarse particles in the rouge.)

A Better Lap. The channeled pitch lap just discussed is the conventional one used by both amateur and professional mirror makers in the past. The one about to be described (Fig. 28) originated at Pennsylvania State College. Introduced into the Amateur Astronomers Association workshop at the Hayden Planetarium by the author, it has been coming into general favor. The amount of effort involved in making the rubber mold may appear to be un-

Fig. 28. The molded lap. At left is the perforated rubber mat centered on the fine-ground face of the mirror, ready for the pitch to be poured. At right is a finished lap.

reasonably out of proportion to the result, but experience has demonstrated that the time spent will pay dividends. Contact is easily made and kept; the lap polishes faster than the channeled lap; and it responds more agreeably to the figuring strokes. The reservoir of water and polishing agent surrounding the facets tends to equalize temperatures and retard evaporation, with the result that turned-down edge rarely occurs. Because of the considerably smaller areas of the individual facets, a somewhat harder pitch than that used for the conventional lap is employed. Less than a third of the usual amount of pitch is needed, and the rubber mold will make innumerable laps.

Sheet rubber, about 7″ square and 1/16″ thick, will do for the mold, the pattern for which is shown in Fig. 29. A saddler's punch, of 3/8″ to 7/16″ diameter, or a No. 4 cork borer, is a good size for making the holes. The radius of the horizontal arcs shown is 24″, but straight lines, rather than curved, might just as well be used. The separations given are for 3/8″ holes, and should be suitably varied for a punch of different size. The pattern is first laid out on a sheet of thin paper, and a 6″ circle then drawn, encompassing as many whole small circles as possible. The pattern is then pasted to the rubber mat with mucilage. When dry, the mat is laid on a thick magazine and the holes are punched out. To cut out the partial holes on the periphery of the 6″ circle, lay the mat over the edge of a board, with the periphery of the circle and the board's edge coincident. By placing the punch carefully over a circle there, an arc of any size can be cut, and the segment of rubber cut out

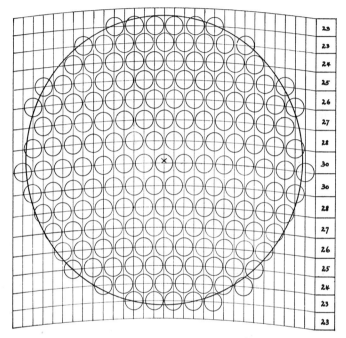

Fig. 29. The pattern of the mat for a 6″ lap, with ⅜″ perforations. Vertical lines are spaced 7/32″ apart. Figures at the right indicate separations between horizontal arcs (straight lines will do as well) in 64ths of an inch.

with a knife or razor blade. Trim the mat to within ¼″ of the holes. Remnants of the paper pattern can be removed by soaking the mat in water.

The operation of making the lap is best done on top of a scrap piece of windowpane or other sheet glass. So that the facets will not squeeze down quickly, the pitch for this lap should be on the hard side. When melted, the pitch should be removed from the stove and allowed to cool briefly, and be thoroughly stirred before pouring. Have the mirror and tool in a pail of hot water while the pitch is melting. Lay the mat on the clean sheet of glass, and paint it all over with a light creamy mixture of rouge[1] and water. While the pitch is cooling, remove the mirror from the hot water and dry it, then place it on the sheet of glass, face up, and paint its surface and sides with rouge. Lay the rouge-coated rubber mat on it, carefully centered. Take the tool from the hot water, dry it and place it nearby, convex side down. Quickly pour the hot pitch over the mat, starting at the center and spiraling outward nearly to the edge. Then, carefully centering the warm tool convex side down over the mirror, set it onto the pitch and press it firmly down to the rubber mat, squeezing the excess pitch out over the sides. After about a minute, invert the disks, and after a similar wait, slide the mirror off. Allowing sufficient time for the pitch to harden, lift the mat from the lap with a gentle stretching and pulling, carefully, so as not to break any edges off the facets. Trim around the edge of the lap with a sharp knife, flush it clean of rouge and pitch fragments, dry it and coat the surfaces of the facets with smoking hot beeswax. All of this takes about 15 minutes, less than half the time and fuss needed for the conventional lap. By working on a sheet of glass, messiness is avoided, and the surplus pitch can be recovered and returned to the pot.

The theory of the graduated spacing (see diagram) is that an oblate spheroid can be brought spherical by concentrating the strokes on the diameter of greatest facet density, and that a spherical mirror can be similarly parabolized. There would be no walking around the barrel, but necessity for rotating the mirror and pressing frequently to maintain a surface of revolution on the lap. But therein lies the rub; a similar surface must be present on the mirror. And so, because of the ever-present danger of an astigmatic surface resulting from this method, the above theoretical practice is not recommended. Used in the normal way, the graduated spacing permits a wider range of stroke than is possible on the channeled lap; thus "zones" are almost entirely eliminated, and the spherical and paraboloidal figures are more easily obtained.

Looking ahead, we find that this mat cannot be used for the lap

---

[1]In this and succeeding discussions, "rouge" refers to any polishing agent.

for the diagonal mirror; the reader may therefore want to consider deviating from the above instructions. (See *The Lap*, in Chapter VII.)

Making Contact. Before polishing can begin, the newly made lap must be immersed in hot water for three or four minutes (five minutes or more for the channeled lap) in order that the pitch may be slightly softened. In the meantime, the barrel top should be fitted with three wooden cleats (Fig. 27); and a clean sheet of paper, with cutouts to fit over the cleats, might be placed on top. Put a spoonful or two of rouge or other polishing agent in a clean jar, add water and stir. Use a small paint brush to apply the mixture (a creamy application for the initial charge) to the face of the mirror. (Watch out for loose bristles.) Remove the lap from the hot water and wedge it firmly between the cleats. Immediately place the charged mirror on top and work it back and forth with pressure for a few seconds in order to embed the polishing compound in the waxed surface of the lap. Then, with the mirror centered on the lap, place a weight of about 20 pounds on top of it, and allow it to press for about 15 minutes, or until all of the heat has gone out of the tool. This process is known as hot-pressing, and should precede each day's polishing operation, the purpose being to bring the surfaces of lap and mirror into contact at every point.

Incongruous results in figuring sometimes occur, traceable to a temporary deformation or bending of the mirror induced by unequal loading during the pressing period. The deformation is impressed into the lap, and therein lies a source of unforeseen trouble. To avoid this and insure equal distribution of the weights, they should rest on a 6″ square or round board which has three pegs, such as rubber-head tacks, driven into it 120° apart in a centrally inscribed circle of $2\frac{1}{8}$″ radius. The pegs, in turn, rest on the mirror.

In putting the lap away for the night, the free polishing compound should be flushed off, and a cover placed over the lap. Nothing, not even the finger tips (save perhaps a clean sheet of paper intended as a cover), should be allowed to come into contact with the surface of the lap.

Plate IV. Polishing the mirror is a delicate operation. Note the position of the arms, and the stance, different from those used in the grinding operation in Plate III.

Chapter V

POLISHING — TESTING — CORRECTING

Theory of Polishing. We know that grinding, as performed by the mirror maker, is a fragmentation of the surface of the glass. The older theory of polishing was that it consisted of a sort of continuation of this process on a fine scale; that is, that the rouge particles, measurable in diameter in units of a wave length of light, were partially embedded in the pitch, the protruding edges having a planing action on the pitted surface of the glass. (This is what occurs when the surface of a pitch lap is charged with a fine grade of carborundum. A fairly good semi-polish can be obtained by charging the lap with fine emery.) Later, the theory was advanced that a molecular flow occurred on the surface glass, induced by a high surface temperature resulting from friction. Substantiating this theory, there is evidence that minute particles of glass become detached from the surface, are adsorbed into the liquid, and either are redistributed or flow away.

In experiments conducted by Dr. E. D. Tillyer,[1] of the American Optical Company, a lens surface polished with rouge was tested and found to contain no trace of iron, but after etching away the surface with hydrofluoric acid, a positive test for iron was obtained, indicating that glass had actually been deposited on top of minute particles of rouge. With further etching away of the lens surface, the actual marks of grinding were revealed, which certainly could not have been there if the polishing action were an abrading one.

In commercial work where close tolerances of figure need not be maintained, the polishing matrix, or rouge carrier, is usually of felt or even of paper. Spectacle lenses, simple magnifiers, and the like are thus polished. In the polishing of lenses, prisms, or mirrors, where high precision of figure is needed, a viscous sub-

---

[1] "Polishing Makes The Lens," N. B. Pope, *Optical Journal & Review of Optometry*, January 1, 1945.

stance such as pitch must be used. The surface of the pitch is charged with the polishing compound which acts upon the surface of the glass. The pitch can be made to conform with the optical surface, contacting it at every point and, in working, will flow so as to permit maintenance of contact with the ever-changing surface of the glass. On large work, channels are cut into the pitch, to facilitate its flow and reduce friction. These channels likewise serve as reservoirs for the liquid mixture of water and polishing agent.

In polishing with either cerium oxide or Barnesite, a higher frictional heat is generated than with rouge, and thereby a more rapid surface flow of glass is probably induced; at any rate, these agents are two or three times faster than rouge. The author can attest to numerous instances in which all trace of scratches that might ordinarily defy rouge was removed in upwards of an hour's polishing with the aforementioned agents. Although there is little choice among the three products in the quality of polish obtained, Barnesite is probably superior in this respect. Some authorities are of the opinion that the fine figuring of optical surfaces cannot be successfully accomplished with the fast polishers, and adhere to the belief that the slower-working rouge should be used for this delicate task. But the drastic action of cerium or Barnesite can be materially subdued by applying it well diluted with water or by the addition of water only, since the charged lap will polish, with lessening effectiveness, indefinitely. So with a dilute mixture, and frequent interruptions of the polishing for long intervals of pressing, figuring can be expeditiously concluded with cerium oxide or Barnesite.

Polishing Procedure. Wax-coated laps sometimes undergo a brief breaking-in period during which the mirror may alternately slip and grab. Thickening the polishing mixture will help to alleviate this condition, although once the lap has been well charged with the polishing agent, it is desirable to work with as watery a mixture as possible.

The usual polishing stroke, which should be continued until all of the pits have been removed from the mirror, consists of a one-third diameter length used center over center, or it can be varied with a W stroke, in which the edges of the mirror are carried out to an overhang of about ½″ on each side of the lap. The broken line in Fig. 30 indicates the path traversed by the center of the mirror in the performance of this stroke. When the lap is in good contact a perceptible drag will be felt, and some effort may be required in stroking. As to pressure, it is difficult to prescribe the exact amount in pounds to be used. A considerable amount may be needed in order to overcome the polishing drag, but it will be less than was used in grinding. Any tendency of the mirror to skid or jump should be resisted. The mirror at all times should pass back and forth across the lap with the two surfaces kept absolutely parallel, and to do this the worker, and not the lap, should control the mirror.

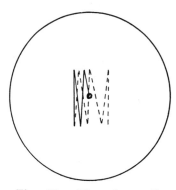

Fig. 30. The zigzag line shows the course traversed by the center of the mirror in executing the one-third diameter polishing stroke.

In order to handle the new technique better, your stance will have to be changed. In grinding, it was possible to stand up close to the work. In polishing, the feet should be spaced one back of the other, some little distance from the barrel. The pivotal action in executing the strokes should start at the feet or hips, not the shoulders or elbows, else there exists a tendency for the mirror to be rocked when the worker is reversing the direction of the stroke, causing a turning down of the edge.

Be sure in polishing that your fingers do not curl over the edge of the mirror to such an extent that they are in close proximity to the concave surface. So placed, they will warm and expand the edge there, which will polish off, and when the mirror has stabilized, its edge will be found to have been turned down. Also, when one's fingers are hooked too far over the edge, an imperceptible tilt may be imparted to the mirror in drawing it back from the end of a stroke, causing its edge to plow into the pitch — result, turned-down edge.

Work at the rate of about 20 to 40 strokes per minute, depending upon the drag of the lap. (Each stroke is the combined back-and-forth motion.) Frequent rotation of the mirror is not necessary, and never rotate it during a stroke, as the excessive action thus given to the edge zones will result in a depressed ring and a turned-down edge. A good procedure is to take six to eight strokes at one position, then shift about 45° to a new position where the strokes are repeated, then another 45° shift, and so on, giving the mirror a slight rotation at each second or third position only.

While the infrequent rotation is perfectly safe in general polishing, and is suggested to avoid excessive edge polishing, in figuring, the mirror should be rotated with each shift of position, as every precaution must then be taken to avoid any possible introduction of astigmatism. This fault results from failure to maintain a perfect surface of revolution. Working for too long a time in one position and taking too many strokes on the same diameter of the lap or of the mirror are obvious causes of producing different radii of curvature on different diameters of the mirror. This is one of the hazards to be contended with in attempting to figure the mirror with a fast-working polishing agent.

After 20 minutes of polishing, wipe the mirror clean with a soft cloth or absorbent tissue, wipe off any lint with the underside of the forearm, and observe your progress. The usual effect of the channeled lap on a spherical mirror is to polish most quickly at the center. A more even polishing action occurs with the use of the molded lap. Imperfect contact may cause zonal polishing. A hyperbolic mirror will polish most rapidly at the edge, in which case the stroke should be lengthened just enough to get the center of the mirror hitting the lap. This will also reduce the hyperboloid. If the polishing falls off abruptly before reaching the mirror's edge, it indicates a flattened or turned-back edge from improper grinding. If, after another period of polishing, there is no indication of the lap hitting this edge (a remote possibility), a return to No. 220 carbo, grinding with the mirror on top and with short strokes, may be necessary.

How to Use the Foucault Device. At this stage you will naturally be curious about the figure on the mirror, and will want to experiment with Foucault's test. It is not essential that this test be conducted in total darkness, although delicate shadow contrasts are then more apparent. Placing the mirror in the testing rack, find the reflection of the lamp window on the piece of ground glass or sheet tin as was previously done in checking the focal length, and by adjusting the rack, bring the sharply focused image to a position like that in Fig. 19. Sit down with your eye about a foot in back of this point and, using a piece of ground glass, locate the position of sharpest focus. Bring the knife-edge up just tangent to this image and in the same plane with it, as nearly as possible, at the same time dispensing with the ground glass. Now reach out and slip the band containing the pinholes down so that the large pinhole is over the window.

If your head is still about a foot back of the knife-edge, the tiny illuminated pinhole image should be seen superimposed on the mirror. By changing the focus of your vision from the mirror's distance to the usual reading distance, you will distinctly see the image of the pinhole suspended in space in line with the mirror and about a foot in front of your eye. The convergent rays from the mirror intersect in this point (see Fig. 21a), then diverge, and it is somewhere in this divergent cone that the eye is now centered. Shift the focus of vision back to the mirror, and now, by bringing your eye forward along the axis of the cone, more of the divergent rays are collected by it, and the pinhole image appears to expand on the mirror's surface until, at a point where both the pupil of the eye and the divergent cone have the same diameter, the mirror appears fully and brilliantly illuminated. The distance of your eye from the knife-edge will then be about $2\frac{1}{2}''$ or more, depending on the size of the pupil. But it would surely try your patience to attempt to test from this position, as your head can hardly be held immobile, and the slightest movement might cause you to lose sight of the image. So your eye should be advanced to within about $1''$ of the knife-edge, where some latitude of motion will be enjoyed with less chance of losing sight of the reflection.

As zonal measurements are not going to be made at this stage, the knife-edge block need not bear against the guide cleat during these preliminary tests. The separation of the pinhole and knife-edge should not be great, however. Theory tells us that if they are at unequal distances from the mirror, an error is introduced in the correction formula for the paraboloid, and if they are widely separated laterally, astigmatism is introduced. But within a range of a few inches either way no measurable error can be observed.

Now push the knife-edge across laterally until it cuts into the rays and causes a shadow to be cast on the mirror. Let us imagine that it is in the position shown in Fig. 31a. Here, it is inside the center of curvature, and the first rays to be cut off are those coming from the left edge of the mirror, so this part of it will darken first,

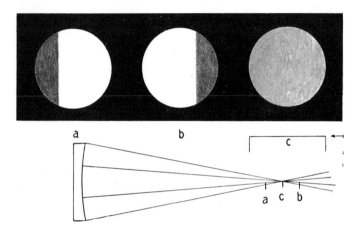

Fig. 31. Knife-edge shadows on a spherical mirror.
a. Inside center of curvature. The shadow moves in from the left, in the same direction as the knife-edge movement.
b. Outside of center of curvature. The shadow comes in from the right, in a direction opposite to the movement of the knife-edge.
c. At center of curvature. The shadow fills the entire mirror evenly and instantly.

the shadow advancing across the mirror from left to right, in the same direction as the knife-edge is moving. Now draw the knife-edge back about ½″ toward the eye, to the position of Fig. 31b. Here, it is beyond the center of curvature, and the first rays to be intercepted come from the right-hand side of the mirror, so the shadow starts there and advances from right to left, in a direction opposite to that of the knife-edge movement.

We have now "bracketed the target," as artillerymen say, and the center of curvature is somewhere between these two settings.

It is finally found by repeated trials, and if the mirror is truly spherical it will darken evenly and instantly the moment the knife-edge has cut completely across the pinhole image, and no direction of shadow approach can be detected. Even with an unduly large pinhole, this condition would still be true.

A spherical surface will hardly be found at this stage, however, and instead of all of the reflected rays intersecting at c, there will probably be a scattering of intersecting zonal rays along a short section of the axis of the cone, as in Figs. 21b and 32, or possibly as in Fig. 39. In these cases, as the knife-edge is cut in, some parts of the mirror will remain illuminated while other parts are in shadow. With a very large pinhole, slight zonal discrepancies, distinguishable only by a faint contrast in shading, might not be apparent in the broad transition from light to shadow, and for this reason a large pinhole is less sensitive than a small one.

It will be difficult to slow down the lateral motion of the knife-edge sufficiently at the moment of its juncture with the axis of the reflected cone of light, so if its motion is arrested as it is about to cut into the rays and gentle downward pressure is applied with the fingers at the right side of the baseboard (Fig. 19), you can probably achieve nice control. If pressing on the baseboard is too lively, apply the pressure to the top of the bench or stand. Always press at the same spot.

In interpreting the shadows which are seen on the mirror, the observer may imagine himself to be in an observation balloon, viewing a *small section* of the earth's surface (the mirror), which may be a plain, a hill, a crater, or any combination of these contours, as it is being illuminated by the grazing rays of the rising or setting sun (located somewhere to the right of the mirror). In this way a sphere, evenly gray in appearance, would resemble the flat surface of a plain. Even without any exercise of imagination, under the shadow test a sphere does look as flat as the proverbial pancake. See Fig. 31c, showing knife-edge setting, shadowgraph, and apparent cross-section of a spherical surface.

An oblate spheroid, with its center zone of longer radius, is shown in Fig. 32. The figure is revealed not only by the knife-edge settings, but by the apparent hill that may be seen on the mirror. The slope facing the imaginary sun is illuminated by it, and the opposite slope is in shadow. (The imaginary hill must not, of course, be presumed to cast a shadow, as the source of illumination is only imagined to be at the right.)   The surrounding plain is gray, and since the knife-edge is at the center of curvature of that zone, it has the appearance of flatness (Fig. 32, b and c).

It is found that different knife-edge settings will produce different shadow appearances and different apparent cross sections of the same mirror.   For example, Fig. 32 shows the shadows and apparent cross sections of an oblate spheroid when the knife-edge is placed at the center of curvature of the center zone *a*, edge zone *b*, and an intermediate zone *c*.   It should be noted that, in the case

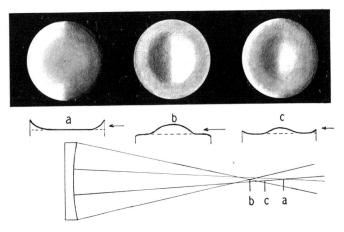

Fig. 32.   Knife-edge shadows on an oblate spheroid.
a. At center of curvature of the center zone.
b. At center of curvature of the edge zone.
c. At center of curvature (not shown) of an intermediate zone.
The shadowgrams represent a typical appearance of an oblate spheroid seen at each of the respective knife-edge settings. Apparent cross sections of the surface are shown directly below. Arrows indicate the direction of imaginary rays presumed to cause the lights and shadows.

of the oblate spheroid, apparent elevations represent glass that lies above that reference sphere (shown by the dotted lines) at the center of which the knife-edge happens to be at the moment.   By removing the elevations from any of these three figures, a sphere would be obtained, but *not*, of course, the *same* sphere in each case.

Fig. 33 shows how the shadows appear on a hyperboloidal mirror, wherein the center zone is of a shorter radius than the edge zone. By means of a diagram, the student should show the knife-edge setting that would produce these shadows.

Further Polishing. One may well be a little slow in comprehending all this, but no matter; get the mirror back on the lap and continue the polishing. You can dwell on the whys and wherefores of shapes and shadows while walking around the barrel. The immediate concern is to polish out the mirror completely, which may take from two to six hours, depending on the thoroughness of the fine grinding, and on the polishing agent used. The polishing mixture should be thinned so that on looking through the overhanging part of the mirror at the end of a stroke, only a faint coloration is visible. Renew the charge only as often as is necessary to keep a supply of liquid in the channels. With the molded lap, a single watery application will last almost indefinitely.

Fig. 33. Shadows on a hyperboloidal 6-inch f/8 mirror. Dotted lines in the lower diagrams represent the reference sphere at the center of which the knife-edge is placed.

After working for a half-hour, give the lap a rest. Center the mirror on it, lay the board on top, and place a weight of about 20 pounds on top of that; then allow to press for about 10 minutes. This operation is known as cold-pressing, and should be indulged in frequently. If polishing spells are too long, the lap may soften and cause trouble. On account of the friction drag, the soft lap tends to follow the mirror as it passes back and forth, causing an excessive deepening of the center. This results in an apparent crater or hole there, or possibly a hyperboloidal figure. A turned-down edge may also result from plowing into the soft pitch. Following numerous periods of polishing and pressing, the pitch may sink a perceptible amount, flowing into the channels and spreading out so that the lap has a larger diameter than the mirror. Therefore an occasional trimming is in order. The channeled lap may be kept as much as 1/16″ less in diameter than the mirror as a safeguard against turning the edge. Don't trim it too small or a turned-up edge may be the result. The harder pitch of the molded lap should stand up long enough to see the mirror through to its completion.

When replacing the mirror on the lap after it has been on the testing stand, you may observe that the polishing drag is not as even as before. This is because the central and more heated area of the lap has sunk more than has the area around the edge, and the mirror will then be riding on a high outer ring of pitch. If polishing were resumed immediately, a turned-down edge and a central hill would result from the mirror being plowed into this ring of pitch. So it is essential, after the mirror has been off the lap for more than a few seconds, that a sufficient amount of time to restore contact (15 minutes or more) be allowed for cold-pressing.

Examining for Pits. Pits are apparent in the early stages of polishing by the grayish appearance of the surface, especially at the edges. This grayness is the diffused reflection from the myriads of tiny pits left by the emery. As polishing progresses it disappears, until the surface has the appearance of being fully polished. But do not be deceived by this first impression. On the silvered or aluminized mirror, the all-but-invisible pits that may still be present would stand out in disastrous prominence. (Some expect the silver or aluminum coating to fill up tiny voids, such as pits and scratches, but as it is only a fraction of a wave length of light in thickness, the coating merely "lines" them and renders them more visible.) With a reading glass, focus the image of the sun or any bright light on the mirror's surface. A small patch of wet paper stuck to the surface will guide the lens to the position of sharp focus. Where any pits are present, the "burning spot" will show up clearly. It can barely be seen on a fully polished surface. Compare its appearance at the center, which area will have the best polish, with that near the edge of the mirror.

Or hold the mirror so that the reflection of an overhead light is seen on its surface, near the edge. Draw it up close to the eye, keeping the reflection in view, and place a magnifier before the eye.

Scan the surface of the mirror at the edge in the vicinity of the light reflection, and the pits can be seen, if there are any, looking like a very fine powdering of dust particles. If you have difficulty in seeing the pits this way, hold the mirror with a frosted lamp back of it and examine the surface with the aid of the magnifier. Any pits should be visible against the luminous background. Polishing should continue until they are no longer visible. If necessary, the stroke should be shortened to facilitate polishing of the edge.

Correcting. Under the Foucault test at the completion of polishing, the figure on the mirror will probably be some variation of the oblate spheroid, perhaps with a turned-down edge. This is because, under the work of fast polishing, the lap has sunk more at the center, and the edge of the mirror has been riding into a higher level of pitch. Sufficient cold-pressing, as discussed above, will have helped to alleviate this condition. If the mirror is hyperboloidal at this stage, either the lap is too soft or too long a stroke was used. Regardless of the shape of the mirror, it should be brought to a spherical figure preliminary to parabolizing.

The first step in the correcting process is to check the condition of the lap. If it appears to be working well and is responding satisfactorily, only a little trimming is in order. If it has spread out around the edges or closed in on the channels, trim it back neatly and carefully with a razor blade. If the surface appears pretty well glazed, scrape off the wax coating with the flat edge of the blade and rewax. If you are using the molded type of lap, leave it alone if it is satisfactory, but if the facets have sunk down or have run together (an indication of too-soft pitch) chip it off with an old chisel, wipe the surface of the tool clean with turpentine, and make a new lap. A dulled and unresponsive lap can often be revitalized by warming its surface over a stove until the wax shows signs of melting, then recharging it and pressing it into contact with the mirror.

Frequently some roughness, zonal rings, or even a pattern of the lap may be seen on the mirror under the knife-edge test. A drastic polishing action or the use of short strokes is the cause of this. Thinning down the polishing mixture and a slight lengthening of the stroke, together with the blending effect of zigzagging (Fig. 30) should smooth up the surface.

Turned-down Edge. This is a narrow zone around the mirror, usually about $1/8''$ or less in width, but sometimes as much as $1/2''$, in which the radius of curvature is rapidly lengthening. On the apparent surface of the mirror, it is somewhat like the run-down heel of a shoe. If left uncorrected in the telescope, a turned-down edge will pitch its rays out beyond the focus of the rest of the mirror, some of them causing an image blur and others passing into the eyepiece and fogging up the field of view.

Under the knife-edge test, there will always be a diffraction ring around the right-hand edge of the mirror, very brilliant if the edge is badly turned. In that case the left-hand edge of the mirror will appear to be soft and dark. If the mirror's edge is perfect, there will be a fine, continuous hairline of light entirely surrounding it, of nearly equal brightness on both sides. The ring on the right should not persist beyond the point where the diffraction disappears from a straightedge suspended across the face of the mirror.

To correct turned-down edge, use short strokes if the edge is badly turned, otherwise the one-third stroke should bring out the diffraction ring, provided proper attention is paid to maintaining contact. Work for five minutes and cold-press for five minutes. After three or four such spells, try the knife-edge test, and if the diffraction ring is making an appearance, continue with that treatment; if not, use a shorter stroke. An excellent aid to securing contact in cold-pressing with the channeled lap (it is not necessary for a molded lap) is the use of a piece of onion sacking. An empty onion bag can be bought at a vegetable store for a few cents. Spread a piece of the material over the lap, lay the mirror on top, and then apply the weights. This breaks up the surface of the lap into numerous small facets which, under polishing, more easily conform to the changing surface of the mirror.

Turned-up Edge. This is a narrow edge zone of shortened radius of curvature. It is conspicuous by a thin strip of shadow just inside the diffraction ring around the right-hand edge. The corresponding zone on the left edge is very bright, and if the zone is very narrow, this may be mistaken for the normal diffraction ring of a perfect edge. Turned-up edge seldom occurs in more than mild form, and usually works out under the long and overhang strokes used to correct the oblate spheroid, or to parabolize.

It may result from the short strokes used to correct a turned-down edge, or from a lap that has too small a diameter. If it is stubborn, polish upside down for a few minutes, with short strokes. This will probably cause the edge to turn down, and some correction will be necessary to restore the diffraction ring. When satisfied with the edge, go after the rest of the figure.

Correcting the Oblate Spheroid. In the discussion of the oblate spheroid, it was stated that any of the three apparent shapes in Fig. 32 could be converted into a sphere. It is obvious that it would be difficult to select a stroke that would correct *a*. The apparent shape at *b* can be handled by the use of a long stroke, but there is quite an amount of glass in that central hill, and excessive long-stroking will turn down the edge. There is less glass in the central hill in *c*, and enough in its edge zones to offset the effects of the long strokes, so that is the knife-edge setting that ought to be used for correction.

Hot-press for contact. Use a one-half diameter W stroke, extending the overhang on each side to the boundaries of the hill. Polish for five minutes and press for 10, and test after two such spells. If the hill is being planed down evenly, continue the same stroke, first restoring contact by hot-pressing if necessary. If a crater or hole is showing up at the center of the hill, reduce the stroke length. If progress seems too slow, lengthen the stroke, or slightly increase the overhang on one side. Watch out for a lengthening radius of the edge zones, an indication that the stroke is too long. Hard laps will permit a wider range of strokes than soft ones. If good contact is being maintained through long periods of cold-pressing, the edge will not be impaired, and the blending overhang stroke will be ironing out any zonal rings or roughness that may have been caused by the short strokes used in edge correction. With the harder pitch of the molded lap, the periods of corrective polishing can be extended to 10 or 15 minutes. Use of a rapid stroke will often bring a hill down fast, but look out for turning the edge. Finally, to wipe out any zonal irregularities remaining from the oblate figure, return to the one-third W stroke (Fig. 30), cold-pressing frequently for long intervals; or hot-press if necessary. Eventually the mirror will be whipped into a spherical figure, and on the Foucault stand it will appear flat and velvety smooth.

Correcting the Hyperboloid. The hyperboloid is usually the
result of overshooting the paraboloid, or too zealous work in reducing the oblate spheroid. In it, the radius of curvature of the
center zone is shorter than that of the edge zone; this is likewise
true, of course, of the ellipsoid and the paraboloid, but with the
hyperboloid the distance between the intersections of these radii
with the axis is greatest. On the Foucault stand, the shadow boundaries on all three figures will be alike, but the shadow *depths* will
differ (see Fig. 33), the shadows darkening at a rapid rate as the
distance between the centers of curvature (*ba*, Fig. 39) increases.
The treatment to be applied here is similar to that described for
correcting the hyperbolic figure that was had from long-stroke
grinding (Fig. 23). There, it will be recalled, the edge zones had
to be worn back by grinding with short strokes. Now, the knife-
edge should be set at the center of curvature of the edge zone, and
the figure seen (Fig. 39a is similar) is taken for correction. The
edge glass must be planed down to the level of the depth at the
center, and this is done with a short stroke. With this stroke, however, the distribution of the polishing cannot be so controlled as to
return the figure to a sphere. Instead, an oblate spheroid is usually
the outcome, and that figure can in turn be dealt with as already
described. Occasionally, where the overcorrection is not great,
and especially with the molded lap, a knife-edge setting similar to
Fig. 39c can be chosen, and the crests, or 70-per-cent zone (explained
in the next chapter), planed down smoothly until the overcorrection
is brought within the allowable tolerance.

It is quite all right to leave the mirror on the lap overnight
if desired, but weights may be dispensed with. Wrap a wet cloth
around the disks to prevent evaporation, and the next day you are
off to a flying start with perfect contact.

Chapter VI

THE PARABOLOID

As MENTIONED in the first chapter, the mirror must have a paraboloidal figure in order that all rays entering the telescope parallel to its axis will converge to meet in a single point in the focal plane (Fig. 10b). It was also stated that a spherical mirror might be altered into that figure in any one of three ways. But not

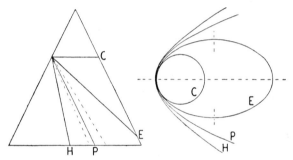

Fig. 34. To obtain the conic sections shown at the right, a cone is cut by planes as drawn at the left: parallel to the base for a circle, C; parallel to a side for a parabola, P; between these for an ellipse, E; in a plane steeper than the parabola for a hyperbola. Rotation of a circle around any axis produces a spheroidal (spherical) surface. Rotation of the parabola and hyperbola about their respective axes produces surfaces of revolution called paraboloid and hyperboloid. Rotation of the ellipse about its major axis produces an ellipsoid; about its minor axis, an oblate spheroid.

the same size of paraboloid would be derived in each case, although, like spheres, the shapes of all paraboloids are the same.

The dotted curves in Fig. 35 represent the paraboloidal surfaces that could be obtained from the same spherical mirror by the three methods. At *a*, the bulk of glass is removed from the central zones, tapering down to zero at the edges. The paraboloid thus obtained has a slightly shorter focal length than the original spherical

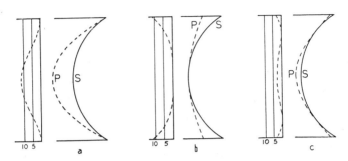

Fig. 35. Three methods whereby a sphere may be altered into a paraboloid. In each case, the graph at the left shows the departure, in millionths of an inch, of the paraboloidal surface from the same 6-inch f/8 mirror.

surface; the new focal length is equal to one half the radius of curvature of the deepened central zone. Of the original sphere, only an extremely narrow edge zone remains as part of the paraboloid. In mirror work, this is the usual method of parabolizing, done by polishing with the mirror face down on the full-sized lap.

In *b*, the bulk of glass is removed from the edge zones of the sphere, tapering down to zero approaching the center, where the original sphere merges with the central zones of the paraboloid. In this case there is no change in focal length, as the central zone may be regarded to have been untouched in the operation. An equal amount of glass is removed by this method, but because the work has to be performed with the lap on top, inevitably leading to turned-down edge, it is seldom attempted.

In *c*, glass is removed from both edge and central zones of the sphere, its 70-per-cent zone (a zone the radius of which is 70 per cent of that of the mirror) remaining as part of that paraboloid. The least amount of glass is removed by this method, which also has to be performed upside down, but with a small polisher, used locally. In this operation, the focal length is lessened by half the amount of reduction made in *a*.

Of course, the mirror sections represented in Fig. 35 are enor-
mously large by comparison with our 6-inch f/8, which occupies so
small a portion of the paraboloidal surface (see Fig. 37) that it
and the sphere are indistinguishable from each other by any ordinary
means of measurement. In the graphs at the left of the diagrams,
however, the departure from the sphere, measured in millionths of
an inch, is scaled to the apparent magnification of about 100,000
obtained with the Foucault test. Note that the differences at a and
b are identical — amounting in the case of a 6-inch f/8 mirror to
11.4 millionths of an inch.

The Reference Spheres. To restate the foregoing, the same
paraboloid can be produced from three different spheres of appropriate radius, as shown in
Fig. 36. The largest sphere,
A, is converted into the
paraboloid by deepening it
at the center. The smallest
sphere, B, is altered by
wearing away glass around
the edge. The center of this
sphere, b, is also the center
of curvature of the central
zone of the paraboloid, the
focal length of which is
equal to $Pb/2$, or PF. By
wearing away glass at both
edge and central zones, the
third sphere, C, is changed
to the paraboloid, its 70-
per-cent zone being all that
remains of that sphere.

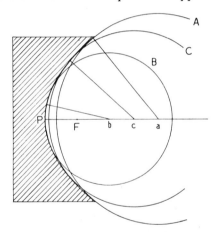

Fig. 36. The three refer-
ence spheres used in testing
a paraboloid.

It is thus seen that, by whatever means a paraboloid is produced, there are three spheres of reference with which the optician
can test the figure of the mirror's surface. There are an infinite
number of other spheres, of course, but the three described are all
that need be reckoned with. The distance between the center of
curvature (b) of the mirror's center zone and the point of intersection with the axis of the normal to any other zone is equal to $r^2/2R$,

where r is the zonal radius, and R is the radius of curvature of the central zone. For a 6-inch f/8 mirror, the value for the edge and central zones is 0.047″ (a and b, Fig. 36), the same as the mirror's sagitta. The center of curvature, c, of the 70-per-cent zone is located exactly midway between points a and b.

Now, by locating the center of curvature of each zone, and measuring the distances between them, we can determine by comparison with the values of $r^2/2R$ how closely the figure of the mirror approaches a paraboloid. In the application of this formula, however, both pinhole and knife-edge would have to be moved together. In practice, it is customary to have the pinhole remain stationary and to move the knife-edge only. In this case, the distances between the points of intersection with the axis (of the rays reflected from the several zones) will have been doubled. The formula then to be applied in determining the length of knife-edge travel is r^2/R.

Although the points of intersection thus obtained (a, b, and c, Fig. 39) no longer mark the centers of curvature of the respective zones, this convenient expression will continue to be used in their connection. Theoretically, the stationary pinhole should be placed exactly at or alongside the center of curvature of the central zone, but no error is introduced as long as it is somewhere in the vicinity (see Chapter V, *How to Use the Foucault Device*). An advantage of thus doubling the knife-edge travel is that the error percentage is halved. The computed value of the r^2/R formula (distance ab, Fig. 39) is known as the *correction* of the paraboloidal mirror, which for the 6-inch f/8 is 0.094″. Ellipsoidal mirrors are referred to as *undercorrected*; hyperboloids are *overcorrected*. The paraboloid is a fully corrected mirror, and is the thin division between the two.

Knife-edge Magnification. The actual separation between the surfaces of a sphere and a paraboloid of equal focal length, at any zone, when placed with their central zones and optical axes coincident, is equal to $r^4/8R^3$, where r is the zonal radius of the mirror, and R is the radius of curvature of the central zone. For the 6-inch f/8, this is 0.0000114″ at the edge (PS, Fig. 37), equal to half a wave length of light. But under the knife-edge test, this departure is seen apparently highly magnified. A group of my students gave estimates of the apparent depressions of the edge zone of a parab-oloidal mirror as seen with the knife-edge placed at the center of curvature of that zone; their values ranged from $1''$ to $1\frac{1}{4}''$. If their average, $1\frac{1}{8}''$, is accepted as a reasonable value, it is found that the Foucault device yields an apparent magnification, of course in depth only, of about 100,000.[1]

Now, at a distance of eight feet (the mirror's radius of curva-ture), a bump or dent of $1/32''$ on a surface is easily seen without optical aid, and its height or depth can be fairly accurately

Fig. 37. Illustrating the relationship be-tween a paraboloid, P, and the reference sphere, S, used in testing, when the knife-edge is placed at the center of curvature, C, of the mirror's central zone. On the scale of the drawing, the 6-inch f/8 is shown by the short heavy line.

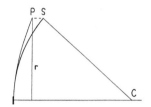

estimated. Assuredly, the presence of a zonal or other surface defect on the mirror, of apparently similar magnitude (the faint marks of the polishing tool that are occasionally seen may be much less), can be detected with the knife-edge and likewise estimated with reasonable accuracy, indicating that errors of the order of three ten-millionths of an inch can be seen.

Surface Tolerance. In 1879 the English physicist, Lord Ray-leigh, showed that for practically perfect image formation all of the light emanating from a point in an object (or from a star) must, after reflection or refraction, meet in a point in the focal plane after traversing paths that do not differ by more than one quarter of a wave length of light. This tolerance is known as the Rayleigh limit. Experience has shown, however, that images quite reasonably approaching the same degree of perfection will be ob-tained if the difference in light path does not exceed twice the Rayleigh limit, or one half of a wave length. (The wave length of

---

[1] This was similarly demonstrated by A. W. Everest, who originated the unique "doughnut" curves for testing the paraboloid. A mathematical discussion of his methods is given by Mr. Everest in the book *Amateur Telescope Making Advanced*.

yellow-green light, which we may adopt as a standard, is 1/45,000″, or 0.000022″.)

In the case of reflection, the light travels twice through the space in front of the mirror. Therefore, the departure of the mirror's surface from a true paraboloid may not exceed a quarter wave in order to meet the more tolerant half-wave limit. To this end, the difference between sphere and paraboloid (distance PS, Fig. 37) must be reduced to this amount, or to 0.0000055″.

The 6-inch f/8 mirror must therefore be corrected (under or over) to within 55/114, or approximately 0.045″, of the value of r^2/R, so that corrections of between 0.05″ and 0.14″ are tolerable. But because of the probable error of observation, it is prudent to have the measured correction lie within 0.07″ and 0.11″. A mirror corrected to the exact value of r^2/R, with a smooth, regular-appearing curve, will perform within the Rayleigh limit, and the finest possible definition will be realized.

Error of Observation. In seeking with the knife-edge the center of curvature of any single zone, which is done through visual interpretation of the shadows cast on the mirror, you cannot, with certainty, repeatedly place the knife-edge within 0.01″ of the true position. A skilled observer can probably halve this error, but with the average amateur, who may make one or at most two mirrors, the error might easily be doubled. Also, assuming that there is no error in the scale, there is nevertheless a possibility of error of 0.01″ in reading it. Of course, where the knife-edge movement is controlled by mechanical means, this latter error should not exist. But as two zones are involved in measuring the mirror's correction, the worker should be prepared to allow for a total error of 0.02″.

Testing the Paraboloid. To provide a means of isolating the various zones that are to be tested, and thus to simplify the problem of locating their centers of curvature, it will be necessary to prepare two masks similar to those shown in Fig. 38. These are made of cardboard, each disk being of the same diameter as the mirror. The one at A has a central opening of about 1½″ diameter, and edge openings about 2″ in length and 0.4″ in width. The openings in the mask at B are ½″ wide (¼″ on each side of the 70-per-cent zone). A good length to make these mask openings is 1½".

The mirror will be tested in three zones; that is, the intersection points of rays from the three zones shown in Fig. 39 are to be sought, and if found where they ought to be, and the curve is smoothly progressive, the mirror may be adjudged a paraboloid.

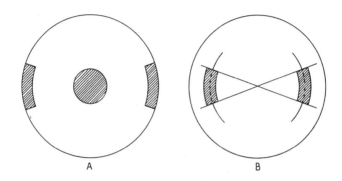

Fig. 38. Masks used in conjunction with the Foucault test
for making zonal measurements on a paraboloidal mirror.

In positioning the testing device preliminary to actual testing, the guide cleat (see Fig. 19) used to guide the longitudinal motion of the knife-edge should, as a matter of convenience, point directly to the vertical diameter of the mirror.

In seeking the center of curvature of any zone, it is a good idea to begin by trying with the knife-edge placed well inside, and then outside of that point — bracketing it, so to speak — and then gradually to close in on it from each side. Knowing how the shadows appear on either side of focus, the observer is able to decide more accurately where the exact center of curvature of the zone lies.

Place the mirror on the testing rack with mask A directly in front of and concentric with it. Set up the testing apparatus, bringing the knife-edge block against the guide cleat so that the indicator is over the scale, and shift the whole device about until after repeated trials the knife-edge is at the center of curvature of the zone exposed by the edge openings of the mask. This occurs when the first faint shadows appear simultaneously in each edge opening, and are at the same time of an equal depth of grayness.

The comparison should be made not when the zone exposures have darkened, but just as the knife-edge encroaches on the rays proceeding from these openings, and while the most delicately gray shadings that permit proper observation are present. This will call for some nice judgment of contrast, particularly as diffraction around the edges of the openings may have a blinding effect (the trick here is not to look at this diffraction), and your eye must dart from one opening to the other to check for simultaneous appearance of the shadows.

The center of curvature thus found is not that of the extreme edge zone, but of a zone of 2.8″ radius, or the mean radius of the zonal opening. A calculation will show that the center of curvature of the extreme edge zone, which is what is ultimately sought, lies about 0.01″ farther from the mirror. Consequently, when it is deemed that the shadow appearances in the edge openings conform to prescription, the knife-edge should be withdrawn from the mirror by that amount. A magnifying glass will be of assistance in making a precise setting. Now remove the mask and observe the over-all shadow appearance on the mirror. As the knife-edge is again cut in, the first shadow to appear will be seen 0.9″ inside the right-hand edge of the mirror (actually at the 70-per-cent zone). With further movement of the knife-edge, this shadow expands, moving somewhat slowly toward the right-hand edge, and more rapidly toward the center of the mirror. Finally, it reaches the mirror's right edge and center together, and at the same moment a thin wisp of shadow appears on the left edge, the mirror then having the dish-like appearance shown in Fig. 39a. As you continue the knife-edge movement, the shadows at the left edge and the center approach and meet at the 70-per-cent zone on that side, and the whole mirror is darkened. These are the shadow appearances when the knife-edge is at the center of curvature of the edge zone, and unless they behave as described, the mirror does not have the paraboloidal shape. Note carefully the indicator reading on the scale.

Replace mask A; then slowly slide the knife-edge toward the mirror, stopping frequently to cut it into the cone of light, and observe the passage of the shadow on the centrally exposed portion of the mirror. At first, the shadow will definitely be seen to come in from the right, because the knife-edge is, relatively speaking, still

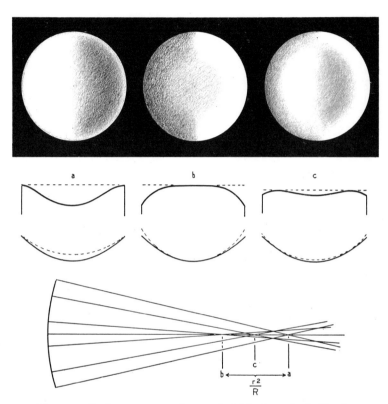

Fig. 39. Shadows seen on the paraboloid when the knife-edge is at the center of curvature of: (a) edge zone; (b) center zone; (c) 70-per-cent zone. Why the shadows behave as described in the text can be found from a study of how, in the lower diagram, the rays proceeding from the various zones of the mirror are intercepted by the knife-edge. Beneath the shadowgrams are shown the apparent cross sections of the mirror seen at the respective knife-edge settings, the broken straight lines representing the reference spheres used in testing, and also the relationship of these spheres with the concave paraboloidal mirror.

well beyond the center of curvature of the central zone, but as this point is approached, the direction from which the shadow comes in becomes less apparent. When a position has been found where this exposed area is seen to darken all over, instantly and evenly, just as though it were a small spherical mirror, and no perceptible direction of shadow approach is apparent, the setting may be accepted as the center of curvature of that zone. Again note the scale reading shown by the indicator. The difference between this setting and the one found for the edge zone should equal the value of r^2/R, or lie within the recommended tolerances.

We must acquaint ourselves with the over-all shadow appearances at all zonal settings, so after removing the mask, we again cut in the knife-edge. The shadow is seen to appear on the left-hand edge of the mirror, moving slowly at first, then with increasing rapidity until upon reaching the 50-per-cent zone (halfway from the center of the mirror to the edge), it suddenly sweeps across the central part of the mirror. The left-hand zones of the mirror are then dark, and the central part, bounded by the mirror's 50-per-cent zone, appears flat and gray. In appearance, the mirror resembles a bowl turned bottom side up (Fig. 39b). Special attention should be paid to the upper and lower shadow boundaries as the shadow encroaches on the mirror. These should appear to terminate at the "poles" of the mirror, from which points the shadow seems to pivot. If there is any straightening or outward bending of the shadow, the indication is that the outer zones of the mirror are insufficiently corrected. This test becomes more delicate if the knife-edge is tried a small distance inside of focus.

It is entirely possible, as we shall see when we come to the parabolizing strokes, for the over-all correction to be right, and yet for the mirror not to have the paraboloidal shape. It is true that the difference cannot be great in the case of the 6-inch f/8; nevertheless the zonal aberration from a mirror of sufficiently irregular figure may be of a rather large order. Of course, at either of the above knife-edge settings, any irregularity in the progression of the curve would be instantly detected by the experienced worker through misbehavior of the shadows, but it might be asking too much of the tyro to expect him to match the veteran's skill in observation and judgment. It is therefore necessary to test an intermediate zone of the mirror, the one most suited to the purpose being the 70-per-cent zone, and for this mask B is used.

The value of the mask for the 70-per-cent zone is questionable, but so many beginners feel that it is beneficial in enabling them better to grasp the fundamentals of the test that its use is suggested.

Now, as closely as possible, place the knife-edge exactly half-way between the above two settings, and cut it in. Faintly gray shadows should appear simultaneously in each zonal opening, there being no indication of the direction of approach. However, if overcorrection is present (or if the mirror has a lesser ratio than f/8), the shadows in each opening may be seen to come in from the left, but they must come in together. If a shadow appears in the left-hand opening first, the indication is that the central part of the mirror is overcorrected; if seen in the right-hand opening first, then there is a preponderance of correction in the edge zones. Here, too, diffraction from the edges of the zonal opening of the mask will have a disturbing effect on the worker in scanning for the shadows, but if these luminous outlines are not *looked* at, their disturbing effect will be materially lessened.

Removing the mask, we once more cut in the knife-edge. The first shadow to be seen moves in on the left-hand edge of the mirror, immediately followed by a second one appearing in a spot midway between the center of the mirror and the right-hand edge, in other words at the 50-per-cent zone. As the knife-edge progresses across the pinhole image, the second shadow expands, spreading quickly to the right, and more slowly to the left; at the same time the first shadow is moving farther in on the mirror.

When the knife-edge has reached the axis of the reflected cone of light (the actual point of intersection of rays from the 70-per-cent zone), the second shadow should extend from the center of the mirror to the 70-per-cent zone on the right; the first shadow should have reached the 70-per-cent zone on the left. The mirror now should have the appearance shown in Fig. 39c. The crests of the convex curves marking the 70-per-cent zone should form the boundaries of the shadows. If the knife-edge movement is now continued, the first and second shadows will approach each other, meeting at the 50-per-cent zone on the left side of the mirror, while the shadow that was stopped at the 70-per-cent zone on the right side will not quite have reached the edge of the mirror there.

Note that in Fig. 39c, the apparently depressed portions of the perfectly corrected mirror, namely, the center and extreme edges, are at the same level. And again note that the 70-per-cent zone marks the crests of the convex curves in the apparent cross section there.

This is the typical appearance of a paraboloid at that knife-edge setting. It is obvious that if the surface of a mirror is thus found to coincide with the paraboloid in the three zones tested, there can be little residual error. Now let us consider the most common cases of mirrors agreeing with the r^2/R formula for the edge zones, but failing to conform at the 70-per-cent zone.

When the knife-edge is at the halfway setting, it may and frequently does happen that instead of the apparent cross section shown in Fig. 39c, shapes like those illustrated in either Figs. 42c or 43c are seen. It is apparent from these shapes that neither of the mirrors represented is a paraboloid, and the centers of curvature of their 70-per-cent zones are displaced. While in either case it is possible to cause the shadow boundaries or apparent crests to land on the 70-per-cent zones by positioning the knife-edge at these displaced centers of curvature, the setting will not be exactly halfway between the centers of curvature of the edge and center zones, nor will the apparently depressed parts of the mirror be at the same apparent level.

Other possible defects are too great a width of the curves at the 70-per-cent zone and insufficient breadth of the central trough, or a condition that is the opposite of this. These errors may escape the observation of the unpracticed eye at the halfway setting of the knife-edge, but their detection should be made possible by the discrepancy in the breadth of the central plateau in Fig. 39b.

Location of the center, 50-per-cent, and 70-per-cent zones should not be left to guesswork. The zones and the center of the mirror can be exactly marked, just prior to testing, with tiny drops of ink deposited on the mirror's surface from the end of a toothpick. When the mirror is set up for testing, the diameter along which the ink-spots lie should be in a horizontal plane; the spots will be conspicuously revealed by diffraction.

In all this shadow testing, no zonal irregularities should be visible; the shadows must flow smoothly from one zone to another. The depth of shading is not great, and so delicate are the transitions from highlight to shadow that it is difficult to reproduce them faithfully in the drawings. A mirror having the same contrasts as depicted in Fig. 39 would be slightly overcorrected, possibly bordering on the maximum allowable tolerance. Every effort must be made to bring the figure within these tolerances, and, in addition, you should strive for the ultimate in performance by bringing your mirror as close to a paraboloid as patience and knife-edge measurement will permit.

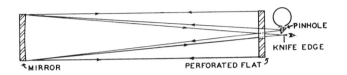

Fig. 40. Testing the paraboloid at its focus with the aid of an aluminized optically flat mirror.

Testing at Focus. A more positive test than the one just described is made at the mirror's focus, using parallel light. The set-up for this test is shown in Fig. 40. The mirror and a silvered or aluminized optical flat are aligned with their axes coincident. (A perforated flat is used in this arrangement, although a small prism or diagonal could be centrally placed before an unperforated flat at a 45° angle, and the pinhole and knife-edge moved around to the side.) Rays from the illuminated pinhole, which is placed exactly in the focal plane, are reflected, parallel, from the paraboloidal mirror to the flat. Thence they return to the mirror, still parallel just as if they had proceeded from a star. The mirror reflects these rays to a point image in the focal plane, and when the knife-edge cuts into that image, the mirror is seen to darken evenly and instantly, just as does a spherical mirror when tested at its center of curvature.

The accuracy of this test is further amplified by the double reflection from the mirror. The flat used must be accurately plane to within 1/10 wave length of light, so it is not a test that the average amateur is likely to employ. But it would be interesting to make the test with your finished telescope on a star, such as Polaris, which is stationary enough for the purpose. Get the star centered in the field, remove the eyepiece, and see if the mirror "blacks out" as a knife-edge is cut slowly across the focal plane.

Test for Astigmatism. In testing for astigmatism, the image of the pinhole should be returned as closely as possible to the source, and in the same plane. This avoids the possibility of any erroneous shadow appearances. If the knife-edge is vertical, any shadow movement at any setting should be in a horizontal direction; that is, if the mirror is considered to be the face of a clock, shadows should approach from the directions of 3 o'clock and 9 o'clock only. Starting with the knife-edge either well inside or well outside of focus, and shifting it gradually and smoothly to a similar distance on the opposite side of focus, observe carefully for shadow approach from any other "o'clock." All shadow movements should be symmetrical about the horizontal diameter. Test on three diameters.

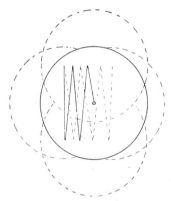

Fig. 41. The zigzag line in this diagram shows the course traveled by the center of the mirror in executing the parabolizing stroke. A similar stroke of less length is useful for reducing a hill in the mirror's figure.

Parabolizing Strokes. The purpose now is to convert the apparent cross section of the sphere (Fig. 31c) into the apparent cross section of a paraboloid, and as this will be done by removing glass from the center of the mirror, the apparent cross section to be arrived at is the one at *a*, Fig. 39. Obviously, if the latter shape is seen on the mirror at some knife-edge setting (the center of curvature of the edge zone), the shapes *b* and *c* will also be found at appropriate settings, and the parabolizing will have been accomplished. Any stroke or combination of strokes that will progressively deepen the mirror into that shape may be used. Before proceeding, the lap should be hot-pressed into absolute contact with the spherical mirror, and about a half-hour allowed before commencing the figuring strokes. A similar procedure should be observed at any stage of the figuring, if the mirror has been removed from the lap for test, or for some other reason.

In general, a narrow zigzagging W stroke gives excellent results. This stroke should carry the mirror from a centered position on the lap to about $1\frac{3}{4}''$ overhang on one side, back across the center to an overhang of $1''$ on the other side, and thence back to center, where the mirror is given a slight rotation, and the sequence repeated from a new position. Too many strokes taken at one position may cause the surface to become astigmatized. After circling the stand once, test the mirror to locate, if possible, the position of the shadow crests. If they are found too close to the center of the mirror, the stroke length or extent of overhang should be reduced. If too far out, the stroke length or the overhang should be increased. The temper of the pitch will determine the extent of stroke length, which may have to be varied somewhat from the nominal figures given above. The harder the lap, the longer the stroke may be. Fig. 41 shows the positions of the mirror at the extremities of the recommended strokes, and the zigzag line is the course traversed by the center of the mirror. The blending action of this stroke is capable of producing a beautiful paraboloidal surface. By proper manipulation, the degree of correction in any one zone can be varied at will, within limits, provided care is taken to prevent disfigurement elsewhere through deformation of the lap.

Too short a stroke will produce apparent shapes like those in Fig. 42. Note that the crests in *c* are too far out. Note also that practically all of the correction lies in the outer zones, the center being flat and undercorrected. Unless this is discovered and the

Fig. 42. Apparent shapes of an imperfectly corrected mirror, seen when the knife-edge is placed: (a) at the center of curvature of the edge zone; (b) at the center of curvature of the central zone; and (c) at the 50-per-cent setting. This surface is the result of using too short a stroke.

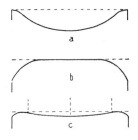

stroke lengthened before the correction amounts to more than about $0.05''$, overcorrection will probably develop by the time the crests have been moved in to the proper position.

If too long a stroke is used, shapes like those in Fig. 43 will result. Here the center zone has received most of the correction. Although knife-edge measurements may show the full correction to be present, the edge zone is undercorrected, or flat, and the

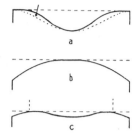

Fig. 43. This diagram is similar to Fig. 42, with the knife-edge settings for the apparent cross sections made at the same locations for each respective apparent shape. This surface is the result of using too long a stroke.

crests are too far in. It is possible to alter this figure to a paraboloid by removing glass from the zone indicated by the arrow at *a*, down to the dotted curve, where the paraboloid is presumed to lie. In attempting this, a stroke of little more than one-third length is used, with the undercorrected zone overhanging the edge of the lap. With local pressure applied at that point, and a blending action introduced by varying the overhang between that point and the mirror's edge, successful correction can often be achieved.

Despite the tyro's best efforts, he will seldom be able to measure the full amount of correction on his mirror, so if the shadow locations are good, it is advisable to call the job finished when the measured correction is about 0.07″ or 0.08″, as there is an excellent probability that the full correction is there anyway.

If the edge has suffered slightly through use of the long strokes, it can be improved or restored by grinding the mirror face down with light pressure on a large piece of plate glass, using No. 305 emery and water for about 15 or 20 seconds. This will grind flat an edge area about 1/40″ wide, and will result in a light loss of about 1½ per cent, but with a beneficial gain in darkness of field. Flush the mirror clean under the tap, dry it, and observe the diffraction ring on the testing stand.

It is seldom that a paraboloid will be achieved on the first attempt. The worker may overshoot the mark and hyperbolize the mirror, or gross zonal errors may occur. If the figure gets out of hand, it should be restored to a spherical shape by methods outlined in the preceding chapter for correcting the hyperboloid. Through failure, which the novice must be prepared to experience, not once, but several times, you are finding out what not to do, and your judgment and skill in figuring are improving with practice. After the third or fourth trial, the figure obtained will be decidedly superior to what you might have accepted on the first try.

Aluminizing. It is best to defer the aluminizing until the cell and other tube parts have been completed; then the mirror and diagonal can be coated at the same time. In Chapter XI, aluminizing and care of the optical parts are treated in detail. Any but a perfect job of aluminizing should be rejected. It is also a good idea to have both primary and secondary mirrors coated with magnesium fluoride or silicon monoxide, giving a hard, durable surface that will withstand much more cleaning than the softer aluminum, without affecting its reflecting quality.

Chapter VII

THE DIAGONAL

RAYS OF LIGHT from a distant star, upon reaching the earth, are sensibly parallel, and the rays that strike the surface of our paraboloidal mirror are contained in a cylinder 6″ in diameter. They are then converged by reflection to form a point image of the star at the focus of the mirror. If a strip of screen is stretched across the focus, and if other stars are present in the field, their images will be observed on the screen extending a number of degrees on either side of the optical axis. The focal surface thus formed is not actually flat, but is a section of a sphere, concave toward the mirror (Fig. 44). But all the star images will not be of equal

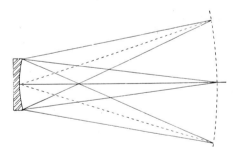

quality, and only the one that lies on the axis will be perfect. Because of the oblique aberrations known as coma and astigmatism (see Chapter XIII), present to a greater or lesser degree in most optical systems, the images deteriorate as they lie farther from the axis, with the result that only a very small part of the field is fit for useful study.

Fig. 44. The mirror can (in the absence of a tube) image a broad area of the sky on the focal surface.

In order to deflect this useful field to one side, so that the observer will not obstruct the incoming rays, a secondary mirror or diagonal is placed a short distance inside of focus, at an angle of 45°. The optical axis is thus deflected by 90°, and the rays from the primary mirror are diverted to form the image in the secondary focal plane (see Fig. 47). The plane reflecting surface of the secondary may also be produced by total internal reflection from a right-angle prism. We shall briefly discuss the merits of a prism and then consider the dimensions of the diagonal mirror called for in the standard design we have adopted.

The objections to a prism diagonal are its greater cost and difficulty of mounting, and the fact that it must necessarily offer a four-cornered obstruction to the light. The glass prism surfaces are easier to keep clean than the aluminized surface of the diagonal, and so are more enduring. In reflective ability both are, ordinarily, about equal. At the 45° angle, the aluminized diagonal reflects about 90 per cent of the light coming from the mirror. Normally, the entrance and exit faces of the prism reflect about four per cent (the hypotenuse face is internally totally reflecting), and there may be a loss of about two per cent due to absorption; about 90 per

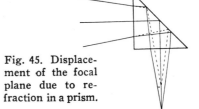

cent of the original light is therefore transmitted by the prism. However, if its entrance and exit faces are given an anti-reflection fluoride coating, the losses there will be reduced to a fraction of one per cent, and there is then

Fig. 45. Displacement of the focal plane due to refraction in a prism.

no question of the superior reflectivity of such a prism.

When a prism is employed, refraction of the converging rays at both of its faces will cause an outward displacement of the focal plane, as shown in Fig. 45. For crown glass of refractive index 1.5, this displacement amounts to one third of the total path length of the axial ray through the prism. (Of course, no axial ray actually enters into the image, as it is obstructed by the prism.)

How Large a Diagonal? Whereas each point image in the primary focal plane would be illuminated by the full aperture of the mirror (in the absence of the tube), only a part of the secondary plane surrounding the axis will receive the benefit of full aperture, the size of the fully illuminated field there being governed by the size of the diagonal. Fig. 46 illustrates the convergence of rays by the mirror to form a single point image on the axis. It can be seen that, wherever the diagonal is placed in that cone of rays, the plane of the cone intersected by its reflecting face is in the shape of an ellipse. As the angle is 45°, the lengths of the minor and major axes of the ellipse may, for all practical purposes, be

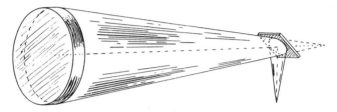

Fig. 46. Convergence of rays from an axial star to a point image at the deflected focus of the mirror.

regarded as in the proportion of one to the square root of two, or 1 to 1.4. The diagonal may be either rectangular in shape, which is simplest to make but needlessly obstructs light and introduces added diffraction, or elliptical. If rectangular, the shape of the diagonal obstruction, when viewed from a point on the axis, is that of a square; if elliptical, the apparent shape is circular.

The need for secondary reflection in the Newtonian telescope has its analogy in the star diagonal of the refractor. This is usually a prism placed a little distance ahead of the focal plane to deflect the image to a more comfortable viewing position. The prism is of such size as to permit the unobstructed passage of rays from the objective to all parts of the largest field to be viewed. This field, measured linearly, is of approximately the same diameter as the field lens of the lowest-power eyepiece. For a starter we might try applying the same reasoning to the problem of the size of the Newtonian diagonal.

In general, the least power that will be used on the reflector will be had from an eyepiece of 1½″ focal length, the field lens of which will be about 1″ in diameter. Accordingly, a field of view of this size, VV' in Fig. 47 (not drawn to scale), is marked off, ½″ on either side of the optical axis, in the primary focal plane FP. The angular size of this field, VMV', can be found as follows: An arc of a circle equal in length to the radius, and called a radian, subtends an angle of 57°.3. As the focal length of the mirror, 48″ for the 6-inch f/8, is the radius here involved, a 1″ length of the arc will contain 1°.2, or 72 minutes of arc. Two stars that are separated by this angular distance may be imaged at *V* and *V′*, and can just be fitted into the field of view of the low-

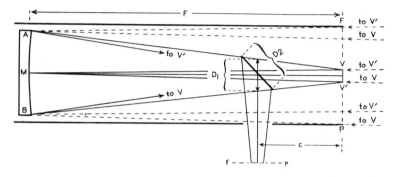

Fig. 47. Schematic diagram of a Newtonian reflector.
The parts are not to the scale of an f/8.

power eyepiece. If the principal rays *MV* and *MV′* have been drawn, only two more rays, *AV* and *BV′*, need be added to complete the representation of image formation of the two stars. (To avoid confusion in the diagram from the use of many lines, rays *BV* and *AV′* are shown in part.) The truncated cone *ABV′V* is thus seen to contain all the rays that flow from the mirror to illuminate every point within the field *VV′*.

The location of the diagonal inside this cone must now be established, and it is obvious that the closer it is to the focal plane the smaller it can be. With a well-designed focusing eye-piece holder, the distance of the secondary focal plane (*fp*) outside the tube can be kept at a minimum, about 1½″. But the amateur, using a telescoping drawtube and a bulky support, will require about 3″, which added to the radius of the tube (next to be established) gives the distance inside of focus for the diagonal.

Now, it is desirable for constructional reasons to have the tube extend to the primary focal plane, where the oblique rays *FA* and *PB*, proceeding from points on the edge of the wide field, are found to be 3½″ distant from the axis of the mirror. Evidently, then, the inside diameter of the tube need be no larger than 7″. But what if it is smaller, as is the case with the aluminum tube shown in the frontispiece? That tube has an outside diameter of 7″, with a wall thickness of 1/16″, and a cork lining ⅛″ thick. It therefore intrudes 3/16″ into those field rays, obstructing a peripheral zone of similar width on the mirror, resulting in an insignificant loss of illumination at the edges of the field of the low-power eyepiece. The major portion of the field, ¾° in extent, is unaffected. With higher-power eyepieces, which may take in a field of this size or less, there is no hindrance by that tube.   Actually, the inside diameter of the tube may be as little as 6½″, but it should not be less than that, as some space must be provided around the mirror to allow ventilation.

The position of the diagonal inside the focal plane is now fixed, in this case, at 6½″ (the radius of the tube plus the distance outside to fp). That is assumed to be the position of the diagonal in Fig. 47, where D_2 is the major axis, and D_1 may be taken for the minor axis. These lengths can be measured from a scale drawing; or the width D_1 is given by the formula,

$$D_1 = \frac{c(M-v)}{F} + v,$$

where v is the linear width of VV', c is the distance of the diagonal from the focal plane, F is the focal length of the mirror, and M its diameter. D_1 multiplied by 1.4 equals D_2. This gives, for the above example, a diagonal of dimensions 1.7″ x 2.4″. But in making the analogy with the star diagonal of a refractor, we did not take into account that the diagonal of the reflector is placed in front of the mirror, forming an obstruction to the incoming light.   So before adopting the above size, consideration should be given to the effects of a central obscuration.

In Fig. 48, which is not drawn to scale, an elliptical diagonal, forming a circular obstruction, is shown 41½″ from a 6-inch f/8 mirror.   Parallel rays from an axial star, and from two other stars each 36 minutes of arc distant from the axis, are shown approaching the mirror.   For the axial star a central area of the mirror equal in size to the area of the obstruction is blocked off. For the field stars, overlapping areas of similar size are also obstructed.   It is seen that a small central area of the mirror, the

black spot in the diagram, will never be used. Furthermore, the
gray zone, a trifle more than 2/5″ wide when the low-power eyepiece
is used, contributes only to the edge of the field of that low-power
eyepiece. In the case of a single star at the edge, less than half
of the gray zone is used.

The black spot is relatively larger for higher-power eyepieces,
and for an eyepiece of ¼″ focal length, which takes in a very much

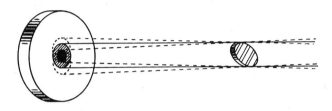

Fig. 48.   Obstruction by the secondary mirror of rays from
axial and off-axis stars.

smaller field, the unserviceable part of the mirror expands to a
diameter only ⅛″ less than that of the gray zone. Thus, the mirror
maker now realizes that he need not be unduly alarmed about
surface defects of any nature in a central area of the mirror of
diameter about equal to the width of his diagonal. This also
explains why there is no loss in perforating the primary when a
Cassegrainian telescope is made.

As the reader may demonstrate to his own satisfaction, these
findings regarding the proportions of the centrally obstructed
areas apply to any size mirror, of any focal ratio, for linear fields
of view of similar diameter (about 1″). Small variations in the
width of the gray zone occur as a result of relative differences in
the distance of the diagonal inside of focus.

Thus it is seen that the larger the diagonal, the less light
there is in the image. More serious, though, is the diffraction that
is introduced. This becomes conspicuous when the area of the obstruction exceeds six per cent of that of the mirror. (See the discussion on diffraction in Chapter XIII.) For these reasons, it
is important that the size of the diagonal be kept at a minimum
compatible with the needs of the observer. The area of the mirror
obstructed by a diagonal measuring 1.7″ x 2.4″ is, if rectangular,

10.3 per cent, and if elliptical, eight per cent of the total. The low-power eyepiece for which this size diagonal is designed is used almost exclusively as a "finder"; when the object sought has been located and centered in the field, the observer invariably switches to an eyepiece of higher power. It is not logical, therefore, to saddle the telescope with a lot of light obstruction and harmful diffraction from a needlessly large diagonal.

The larger diagonal may be justified only in the case of a mirror of low focal ratio, designed for use with low power, where a wide field of view is desired for such observations as the exploration of star fields, comet seeking, and so on. Even there (except for the specific case of variable star work, when the tube, too, should not cut off any part of the field) it is most likely that

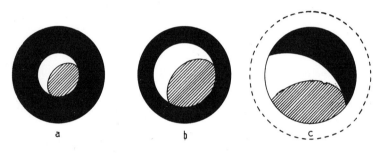

Fig. 49. Increasing the magnification means reducing the size of the field of view. The above illustrations, a, b, and c, show the relative size of the moon's image formed at the focus of a 6-inch f/8 mirror as observed, respectively, through eye-pieces of focal lengths 1½", 1", and ½", the apparent field of each being 36°. The shading indicates the dark part of a crescent moon; the black area is sky in the field of view.

use of a diagonal of, say, 1½" in width, of elliptical shape, and obstructing but six per cent of the mirror's area, would prove far more advantageous than the size prescribed by geometry.

For our f/8 mirror, a minimum practical size may be arrived at from the fact that its field of reasonably good definition is about 30 minutes of arc in width, about the size of the moon. The linear size of the moon's image is found in this way: the tangent of 30′ is 0.0087, which, multiplied by 48 (the focal length of the mirror), gives 0.4176″ or about 2/5″ at the focus. Substituting this value for the width of VV', Fig. 47, the size of the diagonal derived from the formula is then, in even fractions, 1 3/16″ x 1 5/8″. Its area of obstruction, if rectangular, is five per cent, and if elliptical, four per cent of that of the mirror. The amount of light obstructed by it is at a near minimum, and diffraction is reduced to the vanishing point.

The effect on the field of view of this small diagonal in actual observations is illustrated in Fig. 49, where the fully illuminated part of the field, occupied in a and b by the picture of the moon's disk, is shown as it would appear through eyepieces of 1½″, 1″, and ½″ focal lengths, each having an apparent field of 36°. In the first instance, there is a gradual and probably unnoticeable falling off of illumination approaching the edge of the field, so that, for a single star at the extreme edge, the equivalent of slightly less than a 5¼-inch mirror is employed, which is hardly too bad. The losses in the second instance are of insignificant quantity, while in the third instance, the image of the moon (or fully illuminated field, outlined by the dotted circle) is too large to be fitted into the field of view of that eyepiece. If one is situated in a locality where the seeing is generally good, so that high magnifications can be consistently used, a still smaller diagonal might reasonably be chosen, but there appears to be little reason for making it larger.

Preliminary Preparations. The foregoing discussion may answer some of the questions most commonly asked concerning the optics of the telescope which could not be appropriately dealt with elsewhere. And now that the amateur is able to select the size of his diagonal with confidence in his judgment, the question of how to procure it remains to be answered. Indeed, it will be less troublesome and but little more expensive to purchase a diagonal than to make one. But in buying it, the amateur will be denied the fun that comes of making it, and besides, the precision of surface required is not always to be found in a commercial product. The choice of size is restricted, too, to the stock sizes of the manufacturer.

As we shall see, making the diagonal calls for ingenuity and forethought not required in mirror making, and has all the fascination of that activity. But it is neither laborious nor difficult, and once the preparatory operations are concluded, the diagonal itself can be made in a relatively short time. One should be fussy about these preliminary operations, as the departure from flatness must be kept to a very small tolerance.

Some additional equipment is required for making the diagonal, although much of the material gathered for making the mirror will be used. The diagonal will be made of plate glass, which ordinarily comes 1/4″ thick, but the use of plate glass 3/8″ thick is recommended because it offers more resistance to flexure. When cemented to another piece during processing, thin glass may be strained or "sprung" through shrinkage of the cementing agent, particularly if the adjacent surfaces are not mates. As a consequence, after the figured diagonal is freed, the stresses that have been set up in it are relieved, and on return to a normal state the surface is no longer flat.

This effect is readily seen in testing two pieces of glass by the interference method described below. Pressing with the fingers around the edges causes the band pattern to deform and to assume a different shape while the pressure is maintained. When pressure is relaxed, the bands quickly assume a normal shape. Even relatively thick glass will deform under strain, and it will be recalled that in mirror polishing this made it advisable that the weights used in lap pressing be equally distributed on the mirror at three points.

Several pieces of the selected plate glass should be available, or one rather large piece (one or two feet square) which will serve various purposes. A good straightedge; a glass cutter; more abrasives as required, cerium oxide or Barnesite preferred to rouge; onion sacking (already mentioned on page 70); a monochromatic light source; and a setup for testing described below, are among the items to be collected or prepared for making the diagonal. Also, it is important to have another thick piece of plate glass similar to the tool used for polishing the mirror; this tool will also be used.

How Flat a Diagonal? We have taken great pains to produce a curve of remarkable precision on our mirror. We should be equally concerned that nothing will interfere with the formation of a perfect image. In order to keep within the Rayleigh limit, each area of the diagonal that may be engaged in the formation of a point image may not depart from a true plane by more than 1/10 of a wave length of light. Such an area, elliptical in outline, is illustrated in Fig. 46. At a distance of 6½" inside the focus of a 6-inch f/8 mirror, the minor and major axes of this ellipse are 0.8"

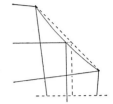

Fig. 50. One effect of convexity in a diagonal. Rays striking its center are projected ahead of their correct positions.

and 1.12", respectively; therefore a diagonal of double these dimensions would have to be flat to within 1/5 of a wave, measured along its major axis. The order of surface precision will, of course, depend on the dimensions of the ellipse shown in Fig. 46, which may vary according to the focal ratio of the mirror and the location of the secondary mirror in the cone. The surfaces of a prism must be flat to a similar degree.

Through examination of Fig. 50, it can be seen that if the diagonal is one wave length convex, the axial ray, aside from any other disturbances to which it may be subject, will be advanced from its proper position by 1.4 wave lengths. Hence, there is need for paying particular attention to the figure of the diagonal. Untested plate glass should never be used. Among several pieces of plate glass, each about 2" square, seldom will one be found that is flat to better than 0.00004". In general, the surfaces of such pieces also have different degrees of curvature on different diameters; in other words, they are astigmatic.

The Light Source. Reasonably monochromatic light is required for testing selected pieces of plate glass as well as the figured diagonal. This light may be obtained from any of the now rather common gas-tube lights; an inexpensive neon lamp is excellent, and argon is also good, but requires a darkened room. Either type may be purchased at a radio supply store. The bulb may be screwed into an ordinary light socket; an adjustable desk lamp is convenient. A thin piece of absorbent tissue wrapped around the bulb will provide satisfactory diffusion. The specimens to be tested should rest on a piece of black cloth almost directly below the light. With these small specimens, the interference bands should be spread out, by pressing, until they are about $\frac{1}{4}''$ apart.

For precision testing of large flats (see Appendix B), the sharper bands to be had from the sodium, helium, or mercury-vapor lamps are much to be preferred. The sodium safelights used in photographic darkrooms are expensive, but ideal for this project. A filter excluding all but the green line should be used with the mercury lamp. With any of these lights a diffusing medium, such as tracing cloth, opal or ground glass, must be used. For accuracy,

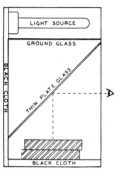

Fig. 51. Using a "beam-splitter" for observing interference bands. By this method, both eye and light source are brought normal to the surfaces under test.

both light source and eye should be placed as near as possible to the normal to the surfaces under test. An arrangement that places both eye and illuminant normal to their centers is shown in Fig. 51. Gross error may result from oblique viewing.

Testing for Flatness. Plane surfaces are tested for flatness by placing two polished surfaces together and observing the interference bands that appear between them when viewed under monochromatic light. The surfaces must be clean and free of all grease and dust or the bands may not show up.

Because of its wave nature, light reflected from the surfaces under test alternately reinforces and interferes with itself, interference occurring with each successive half wave length of separation of the surfaces, which points are marked by the presence of dark bands, or fringes. These bands can be interpreted just as are the contour lines on a topographic map, where the terrain along any given line is all of the same elevation. For example, the surfaces in Fig. 52 are, anywhere along band 2, a half wave length farther apart than they are along band 1; anywhere along band 3 they are a half wave farther apart than along band 2, and so on.

Flat surfaces, *A* in the diagram, or spherical surfaces of equal radii, *B,* present identical band patterns. The direction of greatest separation, or the direction of the wedge opening, can be found by pressing at one side or the other. Pressing at the left tilts the top piece at a steeper angle, widening the wedge opening, and as more bands move in they become thinner and more closely spaced. Pressing at the right, at x, narrows the wedge opening; as the bands move out they become broader and fewer in number until only one band, spread out across the whole specimen, may be seen when the surfaces have been brought parallel to each other.

Where one or both surfaces are not flat, curved bands may be seen, resembling the appearance in Fig. 52C. A little reasoning

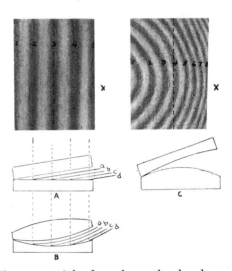

Fig. 52. Testing for flatness, by use of the interference of light waves. It is helpful, in interpreting the meaning of the dark bands, to regard the imaginary surfaces shown by lines a, b, c and d, as being parallel to the surface being tested. The separation between these imaginary surfaces is always equal to one half of a wave length of the light used.

will show that, by placing a straightedge along the bands, as indicated by the dotted line, and counting their number intersected by it, the amount of convexity (or concavity) can be determined. At C, three bands are cut through, hence the surfaces are either convex or concave to each other by $1\frac{1}{2}$ wave lengths. The changing band appearance under appropriate pressure will disclose which is the case. By pressing at the right, at x, narrowing the wedge opening, the bands (in this case) will move off in that direction, and the bull's-eye, or center of the fringe system, will come into view, having the appearance of Fig. 53c (where the same specimens are again tested). The departure from flatness can then be determined by counting the number of rings visible; in this case three rings are present, hence the surfaces are 1½ waves off.

Note that the bull's-eye, in the case of convexity at the point of contact, moves toward the point of pressure. Or, if convexity is

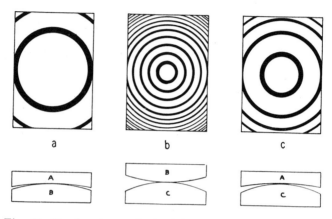

Fig. 53. Band patterns obtained from combinations of three specimen diagonals.

so slight that the bull's-eye is not in evidence, the curved bands, convex toward the wedge opening, move, convex, toward that point when pressure is applied there. In the case of concavity, the bull's-eye would mark the point of widest separation of the surfaces, and in that case, application of pressure at the center would cause the rings to flow inward and to become fewer in number; the bull's-eye would recede from a point of pressure at any edge. With only curved bands showing, as in Fig. 52C, the wedge opening, with concavity, would be at the left; by applying pressure there to narrow the wedge, the bands, concave toward that point, would move in that direction and become fewer in number. By these means, the relationship of the contacting surfaces to one another is easily determined.

In commercial optics, surfaces are tested against a master flat of known precision, so the exact condition of a specimen is immediately known. Although you may have no master flat, you require only three specimens of good plate glass and a little elementary algebra to measure with equal accuracy the flatness of a surface of each specimen. A hypothetical problem is illustrated in Fig. 53, for three pieces of glass marked A, B, and C.

Placing A on B, we get the pattern of bands shown at a, one band convex. B on C gives the pattern at b, eight bands convex. A on C results in the pattern at c, three bands convex. (When making an actual test, several minutes should be allowed to elapse before writing down each result, as heat from the hands will introduce temporary errors.) Convexity is usually assumed to be positive (plus), and concavity negative (minus), so the following simultaneous equations may be written:

$$A + B = +1; \qquad B + C = +8; \qquad A + C = +3.$$

Solve the first equation for B and substitute the result, $1-A$, in the second equation. This gives two equations:

$$A - C = -7, \text{ and } A + C = +3.$$

Solve the first equation above for A and substitute the result, $C-7$, in the last equation, whence $2C = +10$.

From this solution, $C = +5$, $B = +3$, and $A = -2$. Remembering that from one band to the next represents a difference in surface separation of $\frac{1}{2}$ wave length, then

$$A = -2 \text{ bands, or } 1 \text{ wave concave,}$$
$$B = +3 \text{ bands, or } 1\frac{1}{2} \text{ waves convex,}$$
$$C = +5 \text{ bands, or } 2\frac{1}{2} \text{ waves convex.}$$

Making the Diagonal. By the testing procedure just described, we find the exact condition of the surfaces of three pieces of plate glass. Let us keep the best (A in the above case) for testing, and undertake to correct one of the others, say piece C. Other pieces of the same thickness should be cut up so that they can be fitted on the flat side of the glass tool on which the mirror was ground, as in Fig. 54. Or, if the preferably smaller sized diagonal, 1 3/16″ x 1 5/8″ (2″ long if it is to be made elliptical), is going to be used, four or five pieces of this size can be grouped at the center and surrounded with "filler" pieces. By this arrangement, upon completion of the work, you will have the choice of the best of several pieces.

As will be seen from Fig. 59, to avoid needless obstruction of light, one edge of the diagonal should be ground off at a 45°

angle. This is done by holding the diagonal between thumb and fingers, at a 45° angle to the piece of scrap glass on which the grinding is to be done. The bulk of glass can be removed with No. 80 carbo, terminating with No. 220. The other edges can be

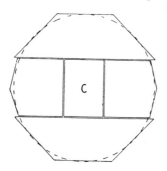

ground squarely and neatly, also with No. 220, and a very small bevel ground on each corner with No. 400. This last operation may best be deferred until polishing and figuring are complete, but then great care must be taken to keep the fingers from contact with the polished surface, or scratches are almost sure to develop. If the arrangement of making several diagonals in the same block is followed, then the grinding of the edges will have to be postponed until final selection of the piece for the diagonal is made. If the

Fig. 54. The diagonal, C, and protective pieces of glass "blocked" on a tool, ready for grinding and polishing.

diagonal is to be made elliptical, the operation described in this paragraph is replaced by that outlined later in this chapter under the section on *An Elliptical Flat*.

The back of the tool on which the mirror was ground often consists of waves and ripples; this also applies to any piece of plate glass, and unless it is first corrected it will not make a satisfactory foundation on which to cement the diagonal or diagonals. In fact, both surfaces must be made to match each other quite exactly, and it is with this aim that we now proceed. The second 6″ disk mentioned among the additional materials needed for the diagonal will be required for the diagonal polishing lap, so by grinding one surface of it and the back of the mirror tool together with No. 400 carbo, the back of the tool can be reasonably well corrected. In order that the tool will not rock on its convex surface when inverted, it should be balanced in a large hole 4½″ in diameter, cut in a thick piece of cardboard.

For this grinding, two charges of carbo should be used, with a one-quarter diameter stroke, alternating the positions of the disks frequently in order to preserve, as nearly as possible, uniform flatness. Next, the several pieces of glass that are to be mounted on the tool (see Fig. 54) should be separately ground, very briefly, against its flat surface with No. 400 carbo. This will bring the surfaces into close harmony, the intention being to eliminate any danger of warping or deformation resulting from the cementing together of ill-fitting surfaces.

Now lay the tool on a board and place it in an open oven, or near a source of radiant heat. Heat slowly and uniformly until the glass is hot enough to melt the paraffin (about 150° F.). The diagonal and the other pieces should also be heated, but not to a degree that they cannot be handled. Remove the tool from the oven, and wipe its flat surface with a stick of paraffin (or brush on melted paraffin), leaving it coated with just a very thin film. Lay the diagonal, ground side down, in the center, and squeeze it down firmly and evenly. Then place each of the other pieces in position, pressing them down also, so that only a very thin film of paraffin binds them. A space of about 1/16" should be left between edges. If, when the assembly is ready for polishing, any abrasive grains lie at the bottom of these spaces and cannot be dislodged, a few drops of shellac will seal them in. Some workers prefer to fill in this space with paraffin or beeswax, but in doing so there is danger of disturbing the seating of the diagonal. If a filler is used, it should then be recessed with the point of a knife, so that it cannot collect abrasive which may later become dislodged and cause scratches. The grooves should be thoroughly scrubbed out when changing grades, and the filler further recessed before polishing.

The over-all surface of the cemented pieces will hardly be found to be in the same plane, but the differences can be corrected by grinding, when cool, against the back or yet unused surface of the second disk. Use No. 400 carbo, a short stroke, and invert positions for each charge in order to forestall the natural tendency toward curvature. By doing the leveling off on the back surface of the second disk, we avoid "grooving" the surface already made nominally flat in the initial operation. To avoid confusion in identity, the first surface should be marked in some way.

Now, the combination "flat" (diagonal and supporting pieces) and the flat surface of the second disk must be ground together with one-quarter strokes, first with No. 600 alundum, and then with No. 305 emery, after which you are ready for polishing. The positions of the disks should be inverted for each successive charge, using eight charges of No. 600, and six of No. 305. When the fine grinding has been completed, both surfaces should test perfectly flat with a good straightedge laid across any diameter.

The Lap. The paraffin-cemented combination cannot be made use of in making the lap, as obviously no heat must be allowed to penetrate it. A channeled lap can be made on the second disk, but a sufficiently large piece of plate glass will have to be used for the preliminary shaping of the surface. For a molded lap, proceed according to the instructions in Chapter IV, making it on the piece of scrap plate glass, which in this case must be substituted for the "mirror," that is, for the combination flat. For flat work, the facets of the molded lap should be uniformly spaced throughout, so another rubber mat will have to be made. If it is thought too much work to make separate mats for mirror and flat, then one of regularly spaced holes, having a 1/16″ tangential separation, should be made to serve for both operations. This will work about as well on the primary mirror as the one with graduated spacing, and it has the advantage of symmetry.

After coating the lap with beeswax, soften it in hot water for a few minutes. The scrap piece of plate glass is once more called into use. Paint it with a thick mixture of the polishing compound, and when the lap is removed from the hot water, work the plate glass vigorously on it, pressing down the uneven wax coating, and charging it well with the polishing agent. Again soften the lap in hot water; then press into contact with the combination flat, first spreading a piece of onion sacking over the lap. Let stand for about 15 minutes, using about 10 to 15 pounds weight in pressing; use the board that was employed in mirror pressing.

Polishing and Testing. For the polishing agent, either cerium oxide or Barnesite is preferred to rouge. After about 15 minutes of polishing, using a one-quarter diameter stroke, a test should be made to determine the benefits of the grinding. The test, of course, should not be made immediately, but after a 15-minute wait. If the diagonal proves to be more than a wave length convex or concave (indicated in the first instance, when tested with hypothetical piece *A*, by straight bands, or curvature due to convexity, and in the second instance, by four or more bands concave), then the difference is best corrected by further grinding.

It is most probable, however, that any error will be within the above tolerances, and polishing can be resumed, the nature of the surface determining the method. If convex, a one-third or longer stroke can be used, in the normal positions; if concave, polishing should be done with the lap on top. The condition of the surrounding pieces does not matter, their purpose being to extend the working surface to such a size that it will not wobble or rock under the motions of grinding and polishing. On account

Fig. 55. Use of a straightedge in testing for flatness. At a, departure from flatness of half a wave is indicated; at b, the difference amounts to less than a quarter of a wave. Dividers should be used for precise measurement.

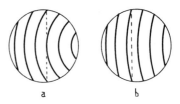

a   b

of the broken-up surface, the onion sacking should be used at all times in pressing. When the diagonal has been completely polished, half an hour should elapse for temperature adjustment before testing.

Try to figure the diagonal as closely as possible to absolute flatness, indicated by testing two bands concave with piece *A*. If an optical flat is available (see Appendix B), curvature of 1/10 of a band is easily measured with straightedge and dividers, as shown above in Fig. 55. Against an optical flat, of course, an appearance like that in Fig. 52, left, indicates absolute flatness.

To free the finished diagonal, place the disk in a pan of very hot water, or lay it on a board and heat it slowly in an open oven. Any wax still adhering to the diagonal can be wiped off with a cloth dipped in turpentine.

If several diagonals have been made in the same block, as previously suggested, it may be desirable to test them one against the other; therefore, in order that the interference bands may be visible, their ground backs should be given a semipolish. This can be quickly done by charging the surface of the lap with No. 600 alundum, and scrubbing the back of each piece on it for a few minutes. Or, if it is desired not to destroy the lap, grind the backs of the separate pieces on a scrap piece of glass briefly with No. 600, followed by a brief polishing on the lap.

The diagonal can now have the final bevel ground on its corners, if the 45° edging has already been done. If the edging has not been done, be careful in the procedure to avoid scratching the polished surface. This operation need not be performed, however, if the diagonal is to be made elliptical.

An Elliptical Flat. With a not excessive amount of labor, a rectangular diagonal can be converted into one of elliptical shape. A simple diagram will show that the length of the rectangular diagonal as prescribed by formula will have to be increased by the thickness of the plate glass used. Therefore, the 1 3/16″ x 1⅝″ diagonal made of ⅜″ glass should be made from a rectangular piece 2″ long, as already mentioned.

A stick of hardwood about a foot in length should be turned up to a diameter equal to the minor axis of the diagonal, and given two or three coats of shellac. One end is cut off at a 45° angle, and the figured surface of the diagonal cemented to it with hard pitch, thus giving it protection in the grinding to follow. In order to make a good bond, both the end of the stick and the diagonal should be preheated before applying the hot pitch.

The corners are ground all around on the scrap piece of plate, using No. 80 for the bulk of the work, and finishing with No. 120. Hold the stick horizontally, and work all over the plate, rotating the stick the while. Five to 10 pounds of pressure can be applied at the diagonal end of the stick. One to two hours will suffice.

To free the diagonal, insert the edge of a razor blade into the pitch binder and strike it a light blow; any pitch adhering to the glass can be dissolved in turpentine. The result is a neat-looking elliptical diagonal, offering a minimum of obstruction to the light. With thick glass, it is most unlikely that its figure will be in any way impaired through removal of the corners.

The above grinding can be done in a lathe as well, holding the stick in the chuck, running at slow speed, and using a sheet of brass or band iron held against the edges of the rotating diagonal, with carbo and kerosene or water fed into it. Particular care must be exercised to protect bearings and ways from abrasive.

There is another way of making the elliptical diagonal, available to the possessor of a drill press. A 6″ glass disk, about 3/8″ thick, and a second one of equal diameter, but at least 3/4″ thick, are required; the tool on which the mirror was ground will also be used. The surfaces of the 3/8″ thick disk, and the one to which it is to be cemented after perforation, will no doubt be at considerable variance with each other, so they should be first ground together, using two charges of No. 400 carbo, working the second one down well. During this operation, the positions of the disks should be alternated several times. This will bring the two surfaces quite flat and into close harmony.

Stand the 3/8″ thick disk on edge diagonally (45° angle) inside a rectangular box 4¾″ or 5″ to the side, and completely surround it with plaster of Paris, filling the box. When it is thoroughly dried out, which may take several days, tack a cover on the exposed side, and remove one of the adjacent sides. With a suitable length of thin-walled brass or copper tubing of correct inside diameter, mounted and held in the drill-press chuck, running at slow speed, and using No. 120 carbo and kerosene or turpentine, bore through the plaster and glass (Fig. 56). Boring will proceed more rapidly if several V-shaped nicks are made in the edge of the tube with a three-cornered file. When the boring is completed, the box can be opened up and the plaster broken loose.

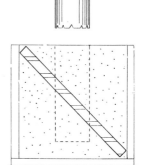

Fig. 56. Cutting an elliptical diagonal from a glass disk.

After perforation, the surface fit between the tool and the diagonal with its surrounding glass disk will have been upset by the relieving of strains in the glass. The surfaces should accordingly again be ground together, separately of course, but this time with No. 600, as the differences will be slight. The perforated disk and elliptical diagonal are now cemented with paraffin to the flat back of the "mirror-tool," in the manner previously described. For surfacing, fine grinding and so on, proceed by methods already described.

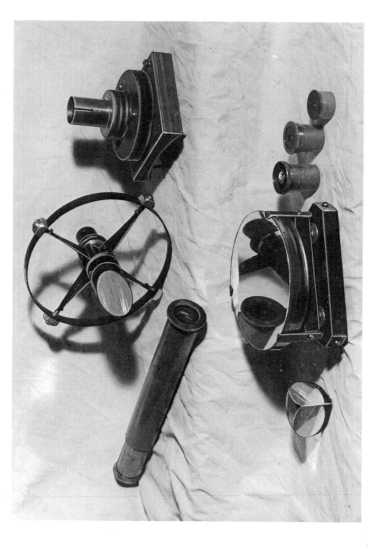

Plate V. Optical parts of a Newtonian: finder, diagonal mirror, spider support, eyepiece adapter tube and holder, prism (alternate for the diagonal), primary mirror in cell, eyepieces.

Chapter VIII

TUBE PARTS — ALIGNMENT — THE FINDER

The Mirror Cell. The wooden cell described here is easily made and permits of ready adjustment. The materials required are: two well-seasoned hardwood boards, one 6″ square and ¾″ thick, and the other about 8″ square and 1″ thick; three ¼-20 round-head stove bolts, 3″ long, and nine washers to fit; three very stiff compression springs, ¼″ to ½″ long and large enough in diameter to fit freely over the stove bolts; three ¼-20 wing nuts; three 10-24 round-head machine screws ⅝″ long; three 10-24's 1″ long. Clips for securing the mirror to the cell can be cut from brass angle shapes obtainable from the 5 & 10.

On the small board inscribe an equilateral "triangle" in a 6″ circle as in Fig. 57a. Mark the holes for the stove bolts 2.1″ from the center. On the large board lay off the equilateral "triangle" inscribed in a circle of the same diameter as the inside of the tube (Fig. 57b). Drill center holes in each board, the size of a dowel stick or bit of rod that may be on hand, and saw out the triangles; the corners must be cut to exactly the proper radius and be made concentric. Center the triangles on each other by passing the dowel stick or rod through the center holes, clamp them, and drill out the stove-bolt holes with a ¼″ drill. Before separating the triangles, mark them so that they can be returned to the same respective positions. Then enlarge the holes in triangle *b* with a 9/32″ or 5/16″ drill to permit free movement over the stove bolts. Drill holes in the angle clips with a No. 12 drill, and elongate them slightly with a small round file. The clips are fastened to the corners of triangle *a* with the ⅝″ screws, first drilling holes for them into the wood with a No. 24 drill. The machine screws will have ample holding power and are preferred to wood screws where frequent removal may be necessary. Before placing the mirror in the assembled cell, file a small flat area

111

on the heads of the stove bolts, and cement to them thin pieces of felt, cork, or leather, on which the mirror will rest. The small dotted circles in the center of each triangle may be cut away with a scroll saw to give ventilation.

The mirror must not be pinched in its cell, as even slight flexure may distort its figure. But if not secured in some manner, it is free to rotate, and occasionally does so from being jarred, throwing its axis out of alignment. To prevent this, a small hole into which the point of a nail or screw may fit should be drilled into the back of the mirror. A nail can be used for a drill. Cut the head off squarely, and insert the pointed end in the drill press. A hole about 1/16″ deep, drilled with No. 120 carbo and water, will be ample. It can be seen from Fig. 57 how the mirror is then anchored. Finally, blacken all interior parts of the cell with a flat black enamel.

The cell is held in the tube by the three 1″ screws, entering triangle *b* through the tube sides. To bring the holes in

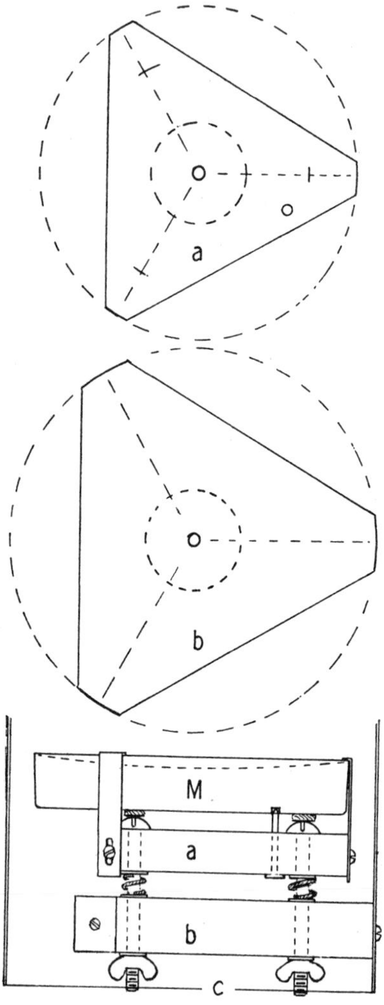

Fig. 57. Mirror cell parts, and assembled cell.

the tube in the same plane, wrap a large sheet of paper around it with the paper's straight edge overlapping in that plane, and mark off the locations of the holes there, 120° apart. After the holes in the tube are made, drill a hole in one corner of the wooden triangle and insert a screw into it through the tube. Steadying the assembly with one hand, bring a second corner opposite one of the other holes in the tube, and mark through the hole on a center line which has previously been inscribed on the triangle. On removing the triangle, this second hole is now drilled, after which the triangle is replaced in the tube, engaged by two screws this time, and the third hole spotted and similarly drilled. In this way, the holes in the tube are transferred accurately to the corners of the triangle.

A refinement which is not essential will provide a much more stable and durable bearing area. After the holes are located in the triangle, drill them out to $1/4''$ size; then insert short lengths of metal rod of that diameter which have first been centrally bored out and tapped for No. 10 screws.

A word of caution is necessary about the making of the wooden triangles. Be very precise about cutting off the corners to the exact diameter, and be sure they are kept concentric, or the mirror cannot be centered in the tube and the efforts to simplify alignment will have been wasted. For easy collimation of the mirror's axis, the cell should be oriented in the tube as shown in Fig. 9, left.

The Tube. This may be made of almost any material, sheet metal, cardboard, plywood, aluminum, or bakelite tubing, or it may be of the skeleton type, although the latter is not recommended except for large telescopes. Rolled sheet metal is perhaps most frequently used. Aluminum ($1/16''$ wall) and bakelite ($1/8''$ wall) are light and accurately round. The cardboard tube around which rugs are rolled is excellent if first doused inside and out with two or three coats of shellac. A tube of any of these materials will provide all the strength and stiffness needed for the telescope.

The tube's length should be about 52", equal to the focal length of the mirror plus whatever is needed to enclose the cell. To locate the eyepiece opening, mark a point on the surface of the tube $41\frac{1}{2}''$ from the plane in which the surface of the mirror

will lie, and with that point as a center describe a circle of ¾"
radius. (The distance of 41½" assumes a focal length of 48",
a tube diameter of 7", and a distance of 3" outside of the tube for
the focal plane. However, see paragraph on page 116 if the tube
is accurately round.) Now drill numerous holes around this circle
with a small drill, not larger than ⅛", and cut out the hole which,
when smoothed up with a file, will be about 1⅝" in diameter.

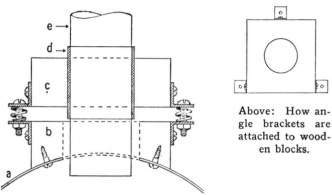

Above: How an-
gle brackets are
attached to wood-
en blocks.

Fig. 58. Adjustable saddle for the eyepiece holder; a, the
telescope tube; b and c, wooden blocks; d, brass bushing; e,
telescoping adapter tube (1¼" inside diameter).

Eyepiece Support. A block of hardwood (maple preferred) 3"
square and 1" thick should then be taken to a pattern maker and
shaped to fit the curve of the tube, and also to have a 1⅝" hole
bored squarely through its center. This block is fastened over the
opening in the side of the tube as shown in Fig. 58. At the same
time have another block of similar dimensions (except see second
paragraph below) bored out to receive a brass bushing having an
inside diameter of 1 5/16". This bushing should be about 1½"
long, with a wall thickness of either 1/16" or ⅛", and must fit
tightly into the block. It should be sealed in with shellac. It will
be much better if the bushing has a flange at one end.

The eyepiece adapter tube, which should telescope snugly and
smoothly into the bushing, is a piece of brass tubing about 4" long,
1¼" inside diameter, with a 1/32" wall. One end should have
three or four longitudinal slots cut into it for a length of about 1";

this can be done with a thin-bladed hack saw. The walls can then be sprung in with slight finger pressure, insuring a firm grip on various eyepieces. Quite often, the adapter tube fits too tightly

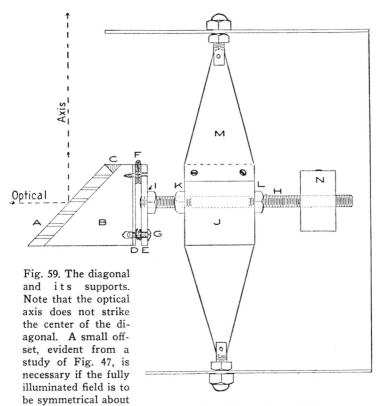

Fig. 59. The diagonal and its supports. Note that the optical axis does not strike the center of the diagonal. A small offset, evident from a study of Fig. 47, is necessary if the fully illuminated field is to be symmetrical about the axis. A, diagonal; B, wood prism; C, wood seat; D, brass plate, 1/16″ thick; E, brass plate, ⅛″ thick; F, hinge; G, compression spring and screw; H, ¼″ rod, threaded at each end; I, lock nut or soldered joint; J, spider support (see Fig. 61); K and L, lock nuts; M, vanes; N, counterpoise.

in the bushing, and some filing and abrading with emery paper must be resorted to in order to effect a smooth fit. A machinist can do this job quickly on a lathe, holding the tubing on a mandrel.

The two wooden blocks are joined as shown in Fig 58, with three compression springs between the matched angles to provide an adjustment that will bring the deflected axis of the mirror and the axis of the adapter tube coincident.

If the tube is accurately round, as, for example, a seamless aluminum tube, it will be possible to dispense with the spring adjustments described. In this case only a single block of wood at least 1¼″ thick is used. One surface is milled to fit the curve of the tube, and a center hole, of a size to accommodate the bushing, is bored radially to this curve, automatically bringing the adapter square to the tube. Even if the adjustment is not quite accurately made mechanically, it can be corrected by inserting shims where needed between the block and the tube. Another advantage of this arrangement is that it makes possible placing the diagonal about ¾″ closer to the focus; as this would alter many of the foregoing dimensions, plan the change carefully if you decide to make it.

Diagonal Holder. At this point a full-scale sectional drawing of the eyepiece end of the tube should be made, similar to Fig. 59, showing in detail the diagonal, holder, spider supports, and so forth. The block on which the diagonal rests, *B*, is a prism-shaped piece of hardwood cut from a square stick of the same width as the diagonal. A ⅜″ square-sectioned strip of wood is glued across the base of the hypotenuse face of *B*, and later one half of it is planed away, making a neat-fitting seat, *C*, for the diagonal *A*. The latter is held in place with three pieces of thin sheet brass (Fig. 60), one of which is tacked to the back and the others to the sides of the wooden prism. These brass pieces should first be annealed by heating them over a flame to a red heat and quenching in water. Use a hardwood stick to bend the "ears" of the pieces over the corners of the diagonal. Do not pinch it tightly—there should be a barely perceptible shake to it.

A 1/16″ thick brass plate, *D*, is fastened to the base of the prism with small wood screws. The small hinge *F* joins that plate to the ⅛″ thick plate *E*; use either small metal screws or solder in attaching *F*. The hinge, which can be purchased at any hardware store, may be so stiff as to resist the action of the compression spring *G*, but it can be completely loosened by applying a pinch of rotten stone or emery and oil to the hinge joint, and working between the fingers for some minutes. A No. 4-36 screw and a compression spring between the plates, opposite the hinge joint, provide adjustment in the angle of deflection.

The plate E screws onto a length of $1/4$-20 rod which has been threaded at both ends. A study of Fig. 47 will disclose the fact

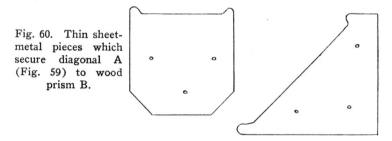

Fig. 60. Thin sheet-metal pieces which secure diagonal A (Fig. 59) to wood prism B.

that the rod, although centered in the tube, should not be centered in the plate, but offset slightly. Failure to take this into account means only that the fully illuminated field of view, circular in outline, will not be concentric about the optical axis, but as the offset in the 6-inch f/8 telescope amounts to only about $1/25''$ it can be ignored. However, if a totally reflecting prism is used in place of the diagonal, it is correct to center it.

The nuts I and K should first have been screwed on, and the joint at I should be soldered. The nuts K and L lock against the central support J, a small longitudinal adjustment being provided to compensate for error in locating the holes for the spider in the tube. N is a small counterweight (that might be omitted) used to remove torque from the vanes.

If ordinary lock washers (not shown in the diagram) are inserted between the nuts K and L and the support J, then after the longitudinal adjustment is effected and the nuts moderately tightened, rotation of the rod H is possible without disturbing the locking mechanism. This is of decided assistance in making alignment of the optical axis with the eyepiece.

Spider Support. Drill a $1/4''$ center hole through a block of hardwood $1\frac{1}{2}''$ long and cut from the same square stick as was the wooden prism B. Saw it out to the shape shown in Fig. 61. For the vanes cut four rectangular pieces, 1½″ x 2⅜″, from sheet metal not less than 0.012″ thick, and attach them with small wood screws to the wood center piece. Holes for attaching the spider to the

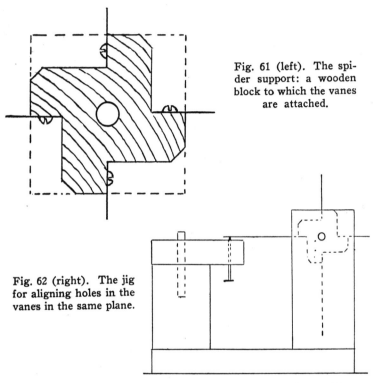

Fig. 61 (left). The spider support: a wooden block to which the vanes are attached.

Fig. 62 (right). The jig for aligning holes in the vanes in the same plane.

tube must be drilled in the vanes so that they will be equidistant from the rotational axis of the diagonal and in a plane at right angles to it. This can be done accurately on the lathe, or the special jig shown in Fig. 62 must be made to insure accurate location. Placed in this jig, with the center rod held rigidly between the two side boards, and the spider rotating in a fixed position, each vane will land in turn on the nail point shown. A light tap with a stick will impress this point in the vane. After drilling out the holes thus marked, the vanes should be trimmed to a triangular shape, as shown in Fig. 59.

For holding the spider in the tube, short lengths of $\frac{1}{4}''$ metal rod are used. Four pieces of this rod, about $\frac{7}{8}''$ long, are threaded for $\frac{5}{8}''$ of their length; a small hole is drilled through the unthreaded end, which is then slotted longitudinally with a hack saw for fastening to the vanes. (Rather than use rod, $1''$ long $\frac{1}{4}$-20 round-head bolts can be bought and their heads sawed off.) Cotter pins may be used to attach these studs to the vanes. The holes in

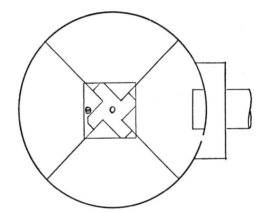

Fig. 63. The position of the spider in the telescope tube.

the tube should be about $9/32''$ in diameter, all four located exactly in the same plane so that the spider will assume the position shown in Fig. 63, thereby giving access to the adjustment screw of the diagonal.

The inside of the telescope tube, adapter tube, vanes, and all other exposed parts should be thoroughly blackened.

The Saddle. Since the eyepiece of the Newtonian reflector is frequently found to be in an awkward position for observing, it is customary to provide a means of rotating the tube in its saddle or cradle (*M* in Fig. 9). A good length for the saddle is about $15''$. The tube rests in V ways, or the ways may be curved to fit the tube. The center of balance of the telescope must be centered on the saddle. To prevent slipping, two rings of metal, plywood, or thick leather belting (*K* in Fig. 9) are secured to the tube, flanking the ways; straps of leather or spring brass (*L* in Fig. 9) hold the telescope to the saddle. A method of attaching bolts to

the straps is shown in Fig. 64; metal straps may be riveted directly
to the bolts. The bolts enter slotted metal bars fastened to the
bottom of the wooden saddle, and extending out about 1″ on either
side (see Fig. 9). To reduce friction in rotating, the straps and
saddle ways should be lined with felt where they bear against the
tube. This material can be held in place with glue or shellac.

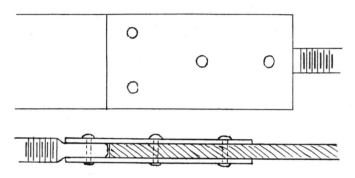

Fig. 64. Joining bolts to the leather (or metal) straps for
securing the telescope to the saddle. (See frontispiece and
Fig. 9.) Ends of the bolts and straps are riveted between
two thin metal plates.

With such an arrangement, rotation is easy, and when it is time to
close up for the night, loosening of the wing nuts allows the straps
to be quickly slipped off so the telescope can be removed indoors.

Optical Alignment. Strap the empty tube in its saddle and
clamp that to the workbench so that it cannot be moved. Set up
a lamp or other light source about 48″ from the eyepiece end of
the tube, and on its extended axis. Then stand a white cardboard
screen, having a ⅛″ hole in it, directly in front of the lamp. Make
two cardboard disks, each with a ⅛″ center hole, and fit one into
each end of the tube. By sighting through the holes, adjust screen
and lamp to bring all three holes in line.

Into each end of the adapter tube fit a wooden plug with a
⅛″ center hole. Then a length of ⅛″ rod is inserted through the
holes in the plugs. Square the rod exactly with the tube longitu-

dinally by means of the spring adjustments on the blocks, checking with a square and rule or dividers. Then push the rod in until it has gone past the center of the telescope tube, and, by sighting through the end of the tube, see if the rod has cut off the light. The adapter must be adjusted until the light is cut off by the rod, then again square the rod with the tube longitudinally, as this first adjustment will no doubt have been disturbed.

Repeat the tests until you are certain that the adapter is square in both planes, after which remove the rod and plugs from the adapter, and also take the cardboard disk from that end of the telescope tube. Now put in the spider, being careful not to disturb the position of the tube, and center it by sighting through the other end of the tube. When the spider is centered, remove the diaphragm from the mirror end and put in the cell containing the mirror. An image of the hole in the cardboard screen will probably be seen somewhere on the screen, and by manipulating the adjusting nuts on the cell make this image fall directly back on the hole. The axes of tube and mirror have now been made coincident and at right angles to the axis of the adapter tube.

All that remains is to insert the diagonal and to bring the deflected axis of the mirror coincident with the axis of the adapter. Fix a small diaphragm of some sort, having a 1/16" hole at its center, over the eyepiece end of the adapter. Placing your eye close to this opening, you will see the walls of the adapter tube, a reflection of the mirror in the diagonal, and in this a reflection of the spider and diagonal, and also a reflection of the hole through which you are peering. These reflections must all be brought concentric with each other, and if the preliminary work has been carefully done, the three adjustments that are provided on the diagonal will be sufficient for the purpose. This last operation is most easily performed outdoors with the telescope pointed at the daylight sky.

The Finder. Usually attached to the reflecting telescope is a small refractor, of short focal length and low power and possessing a wide angular field of view. For a 6-inch telescope, the finder may be anywhere from 1" to 2" in aperture, and 6" to 12" in focal length. When it is aligned axially with the mirror, the sighting of celestial objects is made easy by reason of the finder's large field.

Crosshairs or wires are usually stretched diametrically in front of the field lens of the finder eyepiece, exactly in the focal plane of the objective and intersecting at right angles in the center of the field of view. When the object sought has been brought to the intersection of the crosshairs in the finder, it will then also be in the center of the field of view of the mirror. Instead of crosshairs, a reticle, consisting of a thin glass disk on which cross lines have been etched, may be more simply installed.

Fig.. 65 is representative of the small telescope that is used. From the stores of salvaged war surplus goods, achromatic objective lenses of 1″ to 2″ aperture and 6″ to 12″ focal length can be picked up at very low cost, as can the lenses for the finder eyepiece.

More simply, though, as there is little need to be critical of aberrations, the objective lens might be an ordinary convex spectacle lens and the eyepiece a plano-convex lens of 1″ to 2″ focal length. But as a field at least 3° or 4° in diameter is desirable, a field lens (see discussion in next chapter) of similar focal length should be placed in or near to the focal plane of the objective lens. Means of focusing the eyepiece can be dispensed with, the lenses being secured in a single tube, a la Galileo, with the eye lens fixed at best focus for a star.

Perhaps the most suitable location for a finder is that occupied on the 8-inch telescope in 'Fig. 96, where the two eyepieces are in proximity. Often a star diagonal (right-angle prism) is incorporated in the body of the finder, deflecting the optical axis of its objective lens at right angles, so that both eyepieces may be brought adjacent to each other. In this way the observer looks in the same direction in using either instrument, but opinion seems to have it that it is preferable for the purpose of finding to look in the direction of the object. (In the telescope in Fig. 96, the weight of the eyepiece holder and the finder, both made of heavy bronze parts, necessitated the addition of the counterweight on the opposite side of the tube, to restore the center of gravity to the tube's axis. A sliding adjustment of the weight was provided to compensate for slight error in locating the balance about the declination axis.)

So that the finder can be aligned axially with the main telescope, it is usual to suspend it inside of two rings, and clamp it there by means of screws, in a manner similar to that in which the shafting is shown centered in the fitting in Fig. 80. The rings can be cut from large-diameter tubing, and posts for attaching to the main telescope soldered to them. The minimum height of the finder's axis above the telescope tube should be about 2″.

While the finder is a convenience, it is by no means an essential accessory on a telescope of the size we are making, especially if setting circles are to be attached to the mounting. But if the instrument is to be portable, circles will have little practical value, and should be omitted. Then the telescope may be aimed as though it were a gun, at the approximate spot in the heavens where it is desired to observe; a brief search in that vicinity will usually bring the object sought into the field of a low-power eyepiece. A pair of sights, such as axially aligned protruberances on the tube, may facilitate the above procedure. However, many observers will appreciate the usefulness of a finder, and although it is desirable to encumber a portable telescope as little as possible, the added weight of a small refractor need not be objectionable.

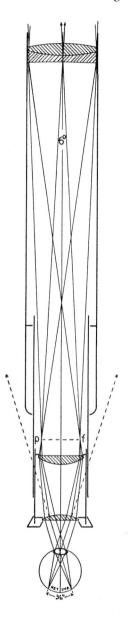

Fig. 65. Plan (about 4/9 actual size) of image formation in telescope and eye. The refractor consists of a 1.7″ objective, focal length 9″, and a Ramsden eyepiece of 1½″ focal length, giving an apparent field of 36°. Magnification is 6x; the exit pupil is 7 mm. in diameter; and the real field of view is 6°. The latter is rather more than is desired in a finder, and for this purpose the focal plane (pf) might be stopped down to about half its present size, or an eyepiece of shorter focal length can be substituted.

Chapter IX

EYEPIECES AND RELATED PROBLEMS

NEARLY EVERY OPTICAL PRINCIPLE as applied to a telescope requires consideration of the function of the eyepiece when the instrument is used visually. Most first telescopes are built for visual observing, so a knowledge of eyepieces is essential if the best possible results are to be obtained from the instrument as a whole. As will be seen in this chapter, there are good and poor eyepieces, and some good ones are not applicable to every type and size of telescope; be very sure, therefore, to equip your instrument with suitable eyepieces of good quality.

Magnification. In general, magnification consists of increasing the visual angular size of an object, which increase can be accomplished by reducing the distance between the eye and the object, either actually or apparently. In Fig. 66a, the distant object D subtends an angular size, a, at the eye. When its distance is reduced by half, as at D', thereby doubling its angular size at the eye, the object appears twice as large (linearly) as before. And so, by means of further angular enlargement, an increasing amount of resolvable detail in the object is made visible.

But there is a physiological limit to which the distance from eye to object can be reduced. For a normal eye, 10″ is accepted as being the distance of most distinct vision; any lesser distance imposes an undue strain on the eye muscles. In Fig. 66b, an object, O, removed about 10″ from the eye, subtends there the angle a. But, the better to define small detail, it is desired to bring the object to within an inch of the eye, at which relative distance its angular size will be increased up to 10 times (for small angles). The eye, however, although strained to the utmost, cannot accommodate for so small a distance, and the object appears blurred.

Therefore, to relieve the strain, a convex lens is introduced

(Fig. 66c) to produce a virtual image of the object about 10″ in front of the eye, where the image can be examined in comfort. To meet these conditions, the lens in this case must have a focal

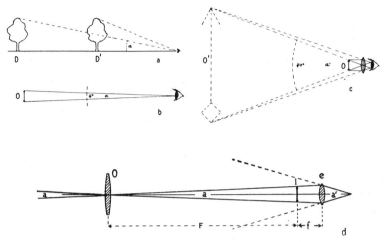

Fig. 66. Illustrating angular magnification.

length of 1″; the image O' will then be, apparently, 10 times the size of the object O, or as the ratio of angle a' to a. It follows, therefore, that the magnifying power of any single lens or combination of lenses *used as a simple microscope* is given by: $M = 10/f$, where f, in inches, is the focal length of the lens, or the equivalent focal length of a set of lenses.

In the case of a very distant object, the aid of a more complex optical system — a telescope — must be obtained. Here a lens or mirror of comparatively long focal length is made to produce a real image of the object at its focus, and the image is then magnified by a second lens (eyepiece) as was the object in Fig. 66c. These are the simple optics of the telescope.

In Fig. 66d, an image of a distant object is formed at I by the objective lens O. The angular size of the distant object at O (or at the eye, since the object distance is very great) is equal to angle a, which is also the angular size of the image I. This image is magnified by means of an eyepiece, e, to the apparent angular size a'.

It is evident that the linear size of the image depends on the focal length of the objective, and also that the focal length of the eyepiece will determine the size of the angle a'. Thus, the magnifying power of a telescope is given by:

$$M = F/f,$$

where F is the focal length of the objective, and f is the focal length of the eyepiece.

Although discussions of magnifying power found elsewhere may differ in some respects from the above, the equation last given is the definition of magnifying power which concerns the telescope when used for astronomical purposes.

Field of View. The *real* (angular) field of view, represented by angle a, Fig. 66d, is measured by that total portion of the image "plane" formed by the objective which can be accepted by the eyepiece. When this field is magnified to the apparent size of angle a', it becomes the *apparent* (angular) field of view. Evidently, then,

$$Real\ field = \frac{Apparent\ field}{Magnification}.$$

It follows, therefore, that as the magnification is increased the field of view becomes smaller. This effect is aptly illustrated in Fig. 49.

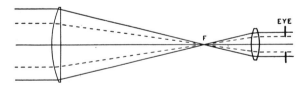

Fig. 67. A simplified diagram illustrating the reduction of aperture resulting from an exit pupil larger than the pupil of the eye.

Exit Pupil and Eye Relief. In Fig. 67, a cylinder of essentially parallel rays from a star is collected by the objective and converged to a point intersection at the focus, which is also the focus of the eyepiece. Thence the rays diverge to the eyepiece and from this they again emerge parallel, but condensed into a smaller cylinder. The size of this cylinder is determined by the diameter of the objective, and also depends on the focal lengths of the eyepiece and objective. The diameter of the cylinder leaving the eyepiece is also the diameter of the exit pupil.

The location of the exit pupil, which is in reality an image of the objective and is called the *Ramsden disk*, will be a short distance back of the eye lens, at the place where the principal ray[1] again intersects the axis after refraction by the eyepiece (see Fig. 65). As all the rays that pass through the objective to make up the image also pass through the Ramsden disk, it is essential that the lens of the eye itself be placed in that plane for most effective viewing. It is desirable to have as large an eye distance (usually called eye relief) as possible, especially with high powers, else one's eyelashes will be rubbing against the eyepiece. In this respect, a single-lens eyepiece gives greater eye relief than the usual two-lens kind. And because this distance is never very great, spectacle-wearers should remove their glasses while observing and make compensation by refocusing the eyepiece; otherwise the field of view may be greatly curtailed.

The diameters of the exit pupil and of the objective have the same relative proportions as the focal lengths of the eyepiece and the objective, so the magnification of a telescope may also be expressed as:

$$Magnification = \frac{Diameter\ of\ objective}{Diameter\ of\ exit\ pupil}.$$

By focusing the telescope on a very remote object and then withdrawing the eye to a distance of about a foot, the exit pupil will be seen apparently suspended in air just back of the eyepiece. It is more conspicuous if, after focusing, the telescope is pointed at the daytime sky. The diameter of the exit pupil can be quite exactly measured by means of calipers or a fine scale held in the plane of the Ramsden disk. A magnifier will aid in obtaining an exact measurement. The above formula will give the magnification of the telescope, and by the F/f formula an exact determination of the focal length of the eyepiece can be made.

---

[1]The principal ray passes through or is reflected at the center of the objective. In the reflector, the principal ray is obstructed by the secondary mirror.

It is evident that as the magnification is reduced, the exit pupil increases in size. But it should not be permitted to exceed the diameter of the pupillary opening of the eye, for reasons about to be explained. In darkness, the average pupil diameter is between 7½ and 8 millimeters, as determined by several authorities. However, in the presence of illumination from a bright star field, some contraction will take place, and a diameter of 7 mm. has been generally accepted as a standard, although it is true that some eyes are able to receive an exit pupil of 7½ mm. or more. Suppose, for example, an eyepiece of 3″ focal length were used with a 6-inch f/8 mirror, yielding a magnification of 16 and an exit pupil of 9½ mm. As only 7 mm. of this could enter the eye, and as the effective aperture of the mirror is equal to the magnification multiplied by the exit pupil diameter, it is apparent that only about 4½″ of the mirror's diameter will be employed. The outer zones are actually diaphragmed out by the iris of the eye. The net result, indicated by the dotted lines in Fig. 67, is a waste of aperture, accompanied by a loss of illumination.

The value of a 7-mm. exit pupil is that it provides the widest possible field of view commensurate with the proportions of the objective, and is desirable in certain kinds of observations, such as comet seeking. A practical example is the 7 x 50 binocular which is standard equipment in the U. S. Navy. This is the so-called night glass, most useful in the deepening twilight. It is a 7-power glass, with 50-mm. (2-inch) objectives. The exit pupil is therefore 7 mm. in diameter, quite filling the expanding pupillary opening in the dim light.

When taking the first look through his telescope upon completing the alignment, an amateur may have been troubled by the conspicuous shadow of the obstructing diagonal, especially if a low-power eyepiece were used at the time. Of course, he was pleased, and perhaps mystified, to find that the objectionable shadow was not apparent at night. This is because the image of the diagonal, projected on the Ramsden disk, may be fully half as large as the pupillary opening of the eye in broad daylight, whereas it is inconspicuously small by comparison with the greatly expanded pupil at night. Therefore, in daytime or terrestrial observations, high-power eyepieces should be used if this disturbing shadow is to be avoided. Low magnifications can be successfully employed with use of the diaphragm shown in Fig. 87, whereby the central obstruction is eliminated. The loss of aperture in that case is not a matter of great concern, as there is usually an abundance of illumination in the daytime.

Theoretically, the smallest exit pupil which ought to be used is one that will exhaust the full resolving power of the mirror. This occurs when the magnification is high enough to reveal all the detail in an object, or, in other words, when the diffraction disk becomes visible. This condition is reached with a magnification of about 13 per inch of aperture, when the exit pupil becomes about 2 mm. in diameter. Practically, however, as stated in Chapter XIII, in the section *Resolving Power*, it will frequently be necessary to resort to considerably higher magnifications.

Eyepieces. The apparent field of a single convex lens used as an eyepiece is considerably smaller than assumed in Fig. 66d. It can be seen in that diagram that unless the angle *a* is small, the field rays from the edges of the objective, after passing through the edges of the image plane, will miss that eyepiece altogether. And, because of inherent aberrations, the sharply defined portion of the visible field will be further reduced, extending not more than 5° on either side of the axis.

It was chiefly to overcome this defect of the single lens that the compound or two-lens ocular was devised. For example, by placing a convex field lens in the vicinity of *I* (Fig. 66), the divergent rays referred to above can be collected and converged and made to enter the eye lens; thus a considerably larger field of view can be encompassed. This is shown to advantage in Fig. 65, the Ramsden eyepiece there having an apparent field of 36°. Placed so close to the focal plane, the field lens contributes but little to the magnification, most of which is accomplished by the eye lens, but by a suitable choice of the focal lengths and separation of the components, aberrations can be materially reduced. The equivalent focal length (*e. f. l.*) of a combination of two lenses when thus used is equal to that of a single lens that will yield the same magnification, and is:

$$e. f. l. = \frac{f_1 f_2}{f_1 + f_2 - d},$$

where *d* is the separation of the lenses.

The linear size of the real field of view can then be taken as approximately the same as the diameter of the field lens of the eyepiece. For an equivalent focal length of 1″, the clear aperture of the field lens may be about 0.7″; a field of this width at the 48″ focus of a mirror will have an angular size of about 50 minutes of arc. As the magnification realized in this instance will be 48 (F/f), the field will be enlarged to an apparent diameter of 40°; a field of this size is frequently attained in a two-lens eyepiece.

Broadly speaking, astronomical eyepieces or oculars can be classified into two types, negative and positive. In the negative type the image is formed between the lenses, or within them, thereby precluding the use of this eyepiece as a simple magnifier. The negative type is represented chiefly by the eyepiece devised by Huygens. Most prominent of the positive types are the oculars of Ramsden and Kellner. All these eyepieces, for astronomical use, should be mounted in standard-sized tubing of $1\frac{1}{4}″$ outside diameter. As to focal lengths, a wide choice is available to the amateur, ranging from about $1\frac{1}{2}″$ down to $\frac{1}{4}″$ and less. The field lens of the lowest power will be about 1″ in diameter, as compared to $\frac{1}{8}″$ or less for the highest powers. Probably the most frequently used eyepiece is one of 1″ focal length, and good choices to supplement it are focal lengths of 2/3″ and 1/3″, although $\frac{1}{2}″$ and $\frac{1}{4}″$ are more popular. Where a telescope does not have a finder, a low-power eyepiece of about $1\frac{1}{2}″$ focal length is a convenient aid in locating objects.

The *Huygenian* eyepiece (Fig. 68, top) is composed of two plano-convex lenses, of unequal focal lengths, in which the convex surfaces face away from the eye. The field lens is placed inside the focus of the mirror, resulting in a slightly smaller image being formed a little closer to the mirror, the image then being magnified by the eyepiece alone. A diaphragm, excluding all but the useful rays, is placed in the focal plane of the eye lens, about midway between it and the field lens. The well-defined apparent field of view is nearly 40° wide. Because of spherical aberration, the Huygenian does not perform as well as the Ramsden or Kellner types on moderate- or low-ratio telescopes, but when used at f/10 and above it leaves little to be desired.

In the usual design of the Huygenian, the focal lengths of the field and eye lenses are in a ratio of 3 to 1, with a separation of half the sum of the focal lengths. Thus, in an eyepiece of 1″ equivalent focal length, the focal lengths of the field and eye lenses are 2″ and 2/3″ respectively. Their separation is 1 1/3″, measured, approximately, between the convex surfaces.

In the *Ramsden,* or positive ocular (Fig. 68, center), likewise consisting of two plano-convex lenses, the convex surfaces face each other. Well corrected for spherical aberration, this eyepiece performs effectively on all sizes of telescopes, and is preferred to the Huygenian for use on Newtonian reflectors. In its most corrected form, the lenses are of equal focal lengths, and separated by a distance equal to their mutual focal length. But with the lenses so spaced, a scratch or any dust on the field lens is brought into sharp focus and magnified by the eye lens. Nor can a reticle be used with such a combination. Accordingly, it has been found expedient to move the field lens out of focus by bringing the lenses closer together, at the cost of introducing a tolerable amount of color into the outer parts of the field. Best separations vary from two thirds to three quarters of the focal length of either lens. The focal plane lies a short distance in front of the field lens, and the eye relief, somewhat better than in Huygens' eyepiece, is thereby further improved.

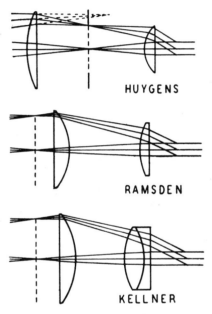

Fig. 68. The principal types of astronomical eyepieces. The dotted vertical lines mark the focal planes.

Specifications for the generally approved Ramsden designs, for equivalent focal lengths of 1″, are as follows: (1) focal length of each lens, 1 1/3″, separation 0.89″; (2) focal length of each lens, 1¼″, separation 0.93″. The lesser spacing affords greater eye relief, and so is preferred for high powers. The good, sharp field of either arrangement is about 30° in extent, but fields from 35° to 40° are usually provided.

A modification of the Ramsden ocular, in which the eye lens is formed of an achromatic combination of crown and flint glass, is called the *Kellner* (Fig. 68, bottom). This eyepiece gives a flat, almost colorless field about 50° in diameter, about as large a field as can ordinarily be used. It is highly recommended in low powers, where its wide field can be used to advantage in variable star work, comet seeking, and so forth.

Specifications, for an equivalent focal length of 1″, are: focal length of field lens, 1¾″; focal length of eye lens, 1 1/3″; separation ¾″. Troublesome "ghost" images are sometimes present with both the Kellner and Ramsden oculars, arising from reflections originating at the convex surface of the field lens and coming to a focus close to the focal plane of the eye lens. These can be materially subdued by coating both field-lens surfaces with an anti-reflection fluoride, at the same time improving the image brightness. Indeed, so marked is the improvement in the performance of fluoride-coated optical elements that uncoated lenses are as outmoded as speculum mirrors.

In a general way, this summarizes the characteristics of the more commonly used astronomical eyepieces. Slight variations from the specifications given are frequently resorted to in order to reduce one or another of the several aberrations. But the amateur must be wary of eyepieces not known to come from a reliable manufacturer of optical goods. A good eyepiece is not cheap — it may cost six or seven dollars or more. Occasionally, an unscrupulous dealer will select from stock an ill-matched pair of lenses that will yield an image, mount them in a cell, and offer them for sale. Such eyepieces are usable, but image formation may suffer considerably, and no reliance can be placed on the assigned focal length.

Aberrations of the Eyepiece. Only the most prominent of these will be briefly discussed: chromatic aberration, distortion, curvature of the field, and spherical aberration. Chromatic aberration is of two kinds, longitudinal (illustrated in Fig. 4), and lateral. The latter is caused by unequal magnification of the different colors, and is more properly known as chromatic difference of magnification. The physical effects of the above aberrations (except spherical) can be readily observed by the possessor of a telescope when a simple lens of about 1″ focal length is employed for the eyepiece. If one of the lenses is removed from a compound ocular, such an experimental eyepiece is had already mounted. (For these experiments, the telescope can consist simply of a small achromatic lens of not less than about 6″ focal length, taped to one end of a short mailing tube.)

Longitudinal chromatic aberration, uniform in amount throughout the field, is revealed when observing a twig or telephone wire against the bright sky. If you push the eyepiece first inside and then outside of focus, the object will be seen in different colors. Both positive types of eyepiece (Kellner and Ramsden) have better correction for this fault than does the Huygens.

The balance of the experiments may be performed by observations on the straight-edged roof top of a distant building silhouetted against a bright sky.

Chromatic difference of magnification is shown by the increasing amount of color that is seen bordering the roof line as it approaches the edge of the field. Axial images are unaffected, but the error increases in intensity in proportion to the distance from the center of the field. The Huygens eyepiece is practically free of this defect, while there may be a troublesome amount in the outer parts of the field of the Ramsden. Some color is also present in the Kellner eyepiece.

Distortion shows up on displacing the roof line toward the edge of the field, when its straight edge is seen to become curved, convex toward the center of the field. This is due to unequal magnification of different parts of the image, the aberration increasing as the cube of the distance from the axis. Intolerable in certain terrestrially used telescopes, this aberration is an innocuous one as far as general astronomical observation is concerned. All three compound eyepiece types are substantially free of this error, although more complete correction is found in the Kellner and Ramsden.

Curvature of the field, increasing in amount with the square of the distance from the axis, becomes apparent when the telescope is shifted so that the roof line again divides the field in half; if the center is in sharp focus, curvature of the field makes it necessary to push the eyepiece inward in order to bring the edge into equally sharp focus. Only partially corrected in the Huygens eyepiece, the defect is almost completely removed from the Kellner and Ramsden.

Spherical aberration (illustrated in Fig. 5) is an insidious error, difficult to detect in a single lens because of the small amount of the aberration present. Where an excessive quantity of spherical aberration is produced, the diffraction pattern (see Chapter XIII) becomes blurred, and contrast suffers. The best obtainable image may remain apparently unchanged in quality throughout a small but perceptible movement of the eyepiece. In any given eyepiece, the fault varies as the square of the angular aperture of the objective with which the eyepiece is employed. (Angular aperture is the angle, at the focus, subtended by the objective lens or mirror — see Fig. 67.) For this reason the Huygens, uncorrected for the defect, does not perform well on telescopes of moderate focal ratio; in fact, at f/7 and lesser ratios, it is entirely unsatisfactory. Both the Ramsden and Kellner eyepieces are relatively free of spherical aberration, with the Ramsden surpassing the Kellner in this regard.

Assembling Eyepiece Components. The making of small lenses is a craft that will not be discussed here, as there is already considerable literature on the subject, available to those who would like to experiment in that direction.[2] It might be mentioned, however, that a spindle and lathe are essential for eyepiece manufacture. Eyepiece making is a fascinating hobby. If one is talented in working to small dimensions, he can, with a little practice and the selection of good glass, succeed in producing as good eyepieces as can be obtained anywhere.

An excellent eyepiece can be assembled from a suitable choice of plano-convex lenses. The focal length of a plano-convex lens is equal to twice the radius of curvature of the convex surface, but as this will very likely be unknown, the focal length can be found by measuring the distance from the lens to the bright image of the sun which is formed by the lens. Or stand it between an illuminated pinhole perforation and a screen. When arranged so that a sharply focused image of the pinhole falls on the screen with the lens exactly

---

[2]"An Introduction to Small Lenses," *Amateur Telescope Making Advanced,* and Orford's *Lens Work for Amateurs.*

midway, the focal length is one half either distance. Or, at any distance, the formula for conjugate foci can be applied:

$$\frac{1}{F} = \frac{1}{D_o} + \frac{1}{D_i},$$

where F is the focal length, D_o and D_i the distances to pinhole and image. Measurements are made from the convex surfaces. The formula for the equivalent focal length of two lenses has already been given. An excellent adjunct to experimentation with small lenses is plasticine or sculptors' modeling clay. A wad of such material stuck on a board will hold the lenses securely fixed, and they can be manipulated at will.

Plate VI. Made by Arthur Kolins, of New York City, this is a pipe-mounted telescope with babbitted bearings, and tripod legs of T-iron shape. The mounting design is shown in Fig. 73.

Chapter X

THE MOUNTING

EVERYONE IS MORE-OR-LESS FAMILIAR with the surveyor's transit, in which two axes at right angles to each other enable the telescope to be aimed at any point in the celestial dome. Its motions are one in azimuth, about the perpendicular axis, and one in altitude, about the horizontal axis. The names of these motions have been combined in the term *altazimuth*, which is used to describe this type of mounting. When the telescope is swung through 360° in azimuth, the field of view traverses an orbit about the zenith, whereas the stars, in their daily or diurnal motion, describe orbits around the celestial pole.

However, if the instrument is tilted so that its perpendicular axis points to the celestial pole, instead of to the zenith, the former azimuth motion of the telescope will exactly follow the paths of the stars. This is then described as a motion in right ascension; its axis of rotation, parallel to the axis of the earth, is called the polar axis. The former altitude motion has now become one in declination, turning on the declination axis. This is now an *equatorial* mounting, taking its name from the great circle basic to this system of co-ordinates, the celestial equator.

When the telescope is mounted in this way, objects too faint to be seen with the eye can easily be found. Suppose, for example, we wish to observe the planet Neptune, assuming it to be above the horizon. We find its position (right ascension and declination) in the *American Ephemeris*. Then we swing the telescope on its declination axis so that it points the required number of degrees north or south of the celestial equator, and clamp it. Next we turn the telescope on its polar axis until the correct hour angle is read on the hour circle, and Neptune should be in the field of view. Of course, this procedure is possible only on an accurately adjusted telescope, equipped with setting circles, and if the sidereal time is known or computed for the time of observation. (See Chapter XII.)

There is one feature that must be emphasized in the design of a telescope mounting, and that is rigidity. Nothing is as abominable and useless as a telescope which trembles in the slightest breeze. Take a stick from a fire and whirl its glowing end before the eye in all sorts of gyrations — that is what the image of a star looks like in a weakly mounted telescope. Remember that any shake or vibration is amplified by the magnification being used, and what may appear to be a quite stable instrument may not prove so when you are trying to observe the Cassini division in Saturn's ring with a $\frac{1}{4}''$ eyepiece. Large size, smoothly operating bearings, well-fitting parts, and a sturdy support are required in an efficient mounting.

There is too great a latitude in ideas and design in telescope mounts for the subject to be entered into here. Our primary objective is to enable the amateur to mount his telescope cheaply and

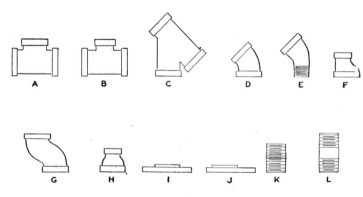

Fig. 69. Useful pipe fittings: A, tee; B, reducing tee; C, wye or lateral; D, 45° elbow; E, 45° street elbow; F, eccentric reducer coupling; G, offset coupling; H, reducer coupling; I, heavy flange; J, offset flange; K and L, close and short nipples.

efficiently. What will be described is a conventional form of mounting known as the German type, utilizing pipe fittings, which are admirably adaptable to the purpose. A study of the problems of the construction will be of aid to those fortunate amateurs who are in a position to design and machine their own castings. The pipe fittings are obtainable in iron or brass at any steamfitters' or plumbers' supply house. Useful fittings are illustrated in Fig. 69.

Shown in Fig. 70 is a fixed pipe mount in which the polar and declination axes turn on the threads of the short nipples *L*. The fittings are of 3″ pipe size, and can be identified from Fig. 69. The height of *A* above the ground should be such that the eyepiece is at a comfortable height when the tube is pointed to the zenith. For better bearing action, the threads of the polar and declination axes should be lapped in. To do this, paint the threads with a mixture of No. 400 carbo and cutting oil, and screw the tee, or flange,

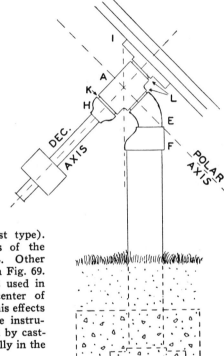

Fig. 70. Pipe mount (first type). The axes turn on threads of the close or short nipples, L. Other parts can be identified from Fig. 69. The eccentric reducer F is used in an effort to bring the center of gravity over the pier. As this effects only partial correction, the instrument's balance is improved by casting the pier base eccentrically in the concrete block.

on and off repeatedly, grinding away until a few more threads become engaged. Then clean out all grit with a stiff brush and kerosene oil, screw the parts together again, and add lubricating oil. A much smoother turning action will be the result. No other machining is required on this simplified mounting. The counterweight may be of iron, lead, or concrete.

Another design is shown in Fig. 71, where a minimum of machining is called for. For this design, not less than 1½" pipe and fittings should be used, and preferably a size larger. The lengths of pipe which make up the polar and declination axes are

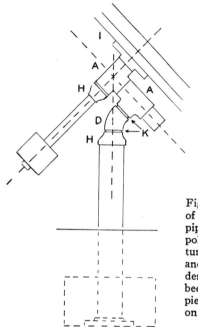

turned perfectly round in the lathe, taking a minimum cut for the purpose. The internal threads of the tees are bored out only enough for the trued-up pipe to be fitted snugly into them. The faces of the tees ought also to be faced off, square to the boring.

Fig. 71. Pipe mount (second type), of which Fig. 93 is an example. The pipe sections which comprise the polar and declination axes have been turned accurately round in the lathe, and rotate in bored-out tees. In this design, the center of gravity has been brought over the center of the pier, and no counterweight is needed on the polar axis. Parts can be identified from Fig. 69.

A third and perhaps the most satisfactory way of employing pipe fittings is shown in Fig. 72. In this design, 1" pipe is used for the axes, with 1½" fittings, and the intervening space is filled with babbitt metal. This gives a very smooth, rigid bearing, and it is possible in the construction to align the axes perpendicular to each other with good precision.

For the permanent mount, where weight is not a factor, 1⅜" solid bar stock is preferred to pipe, which must be machined perfectly round in the lathe. Bar stock is ground truly round, and it is only necessary to have the local plumber cut the pipe thread on it. Be sure in so doing that the vise jaws do not grip that part

TABLE I. SPECIFICATIONS OF PARTS FOR MOUNTINGS*

		Fixed mount	Portable "A"	Portable "B"
A	Heavy flange	1½	Tripod head	Tripod head
B	Lock nuts	½	None	None
C	Wye	1½ x 1½ x 1½	Design of Fig. 71	Design of Fig. 73
			1¼ elbow	2 street elbow
			1¼ tee	1½ x 1½ x 2 tee
D	Close nipple	1½	1¼ (?)	2 (?)
E	Reducing tee	1½ x 1½ x 1	1¼ x 1¼ x 1	1¼ x 1¼ x 1¼
F	Length of pipe	1⅜, 9 long**	1, 8 long	1 & 1¼, 8 long
G	Reducers	1 x ¾	1 x ¾	1 x ¾
H	Extension pipe	1 x 12	¾ x 12	¾ x 12
I	Counterweight	Two	Dec. axis only	Dec. axis only
J	Collars	1⅜	To order	To order
K	Heavy flange	1	1	1
L	Clamping screws	⅜	⅜	⅜
M	Setting circles and pointers	Polar and dec. axes	None	None

* All figures in this table give dimensions in inches.

**Solid ground shafting.

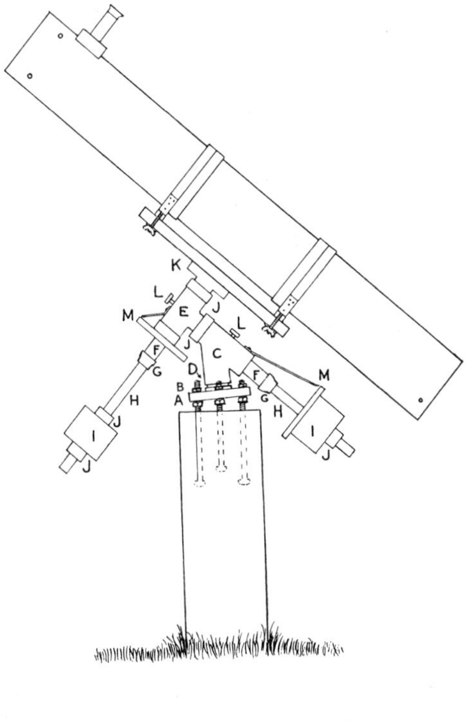

Fig. 72. Babbitted bearing pipe mount. Adjustment of the polar axis for latitude is made with lock nuts B, and a correction in azimuth is made by rotating the wye fitting C on the threads of the close nipple D. See Table I for identification of parts.

of the shafting that is to be encased in babbitt, as the resultant dents will cause trouble. Machined collars to fit the 1⅜″ shafting can be purchased at the hardware store.

It ought not to be difficult to interest a local plumber sufficiently

TABLE II. SPECIFICATIONS OF STANDARD PIPE SIZES*

Size	Outside diameter	Inside diameter	Nominal wall thickness	No. threads per inch	Depth of thread
½	0.840	0.546	0.147	14	0.0571
¾	1.050	0.742	0.154	14	0.0571
1	1.315	0.957	0.179	11½	0.0696
1¼	1.660	1.278	0.191	11½	0.0696
1½	1.900	1.500	0.200	11½	0.0696
2	2.375	1.939	0.218	11½	0.0696
2½	2.875	2.323	0.276	8	0.1
3	3.500	2.900	0.300	8	0.1
3½	4.000	3.364	0.318	8	0.1
4	4.500	3.826	0.337	8	0.1

*All dimensions are in inches. In speaking of pipe size, the dimension referred to is always the approximate inside diameter of *standard weight pipe;* for example, 1″ pipe, whether of standard, heavy, or extra-heavy wall thickness, has the same nominal outside diameter, 1.315″.

for you to borrow the paraphernalia requisite for the babbitting job — a fire pot on which to melt the babbitt, a solder pot, ladle, and torch. Or it is quite feasible to bring the assemblage to his shop and do the job there under his guidance. The entire operation of melting the babbitt and pouring a bearing may not take more than 15 minutes. About three to four pounds of babbitt will be needed for the bearings, but a surplus is needed for the pot, and may be used later in casting a counterweight. Use a good grade of babbitt; the difference in the cost of a few pounds is too small to risk deformation of a bearing that is almost pure lead. The babbitt must be heated to a point where a piece of cardboard will char if thrust into it.

The Babbitted Bearings. First, drill three holes radially near each end of the tee and wye (see Fig. 80) for No. 6-32 screws which will be used to center and support the shafting in the fittings while the babbitt is poured. The ends of these screws should be smoothed with a file so that they will not mar the surface of the

shafting. Holes for the 3/8" clamping screws (L, Fig. 72) should
be drilled and tapped into the tee and wye, and the screws, coated
with lampblack and oil, inserted prior to babbitting and brought
up just to contact the shaft. The ends of these screws should first
be filed quite flat. Two or three V cuts should also have been made
across the internal threads of each of the fittings, the purpose being
to provide grooves that will act as anchorage for the babbitt
bearing.

Make up the polar axis tight in tee E. It will have to be removed later, so in order that it may again be returned to the same
exact position, drill a hole through the branch opening of the tee
into the shafting. In the final assembly, a pin or screw should be
inserted in this hole. Slip the collar into place, snug against the
tee. Now smear the polar axis and the collar with a mixture of
thin lubricating oil and lampblack, to prevent the babbitt sticking
to the metal.

Place the polar axis in the wye fitting, with its collar bearing
against the upper face of the wye, and carefully center it with the
set screws. To prevent the molten babbitt from running out, fit an
asbestos paper or thin sheet-metal collar over the other end and
seal up any gaps with putty. One or two pinhole perforations in
the collar will allow the escape of air. Brace the assemblage so
that it cannot roll during the pouring operation (see Fig. 81).

A poor bearing is apt to result if the babbitt is poured into
cold metal, so while the babbitt is melting a torch should be used
to preheat both casting and shaft. In heating the shaft, apply the
flame to its protruding ends, else the oil film may be burned away,
and there is then danger of the babbitt sticking to the metal. It is
not necessary to bring the parts up to the temperature of the molten
babbitt, which melts at from 500° to 600° F.; about 200° to
250° will suffice.

Make a trough of the asbestos paper or thin sheet metal,
support it in the branch opening of the wye so that the babbitt,
when poured, will not land directly on the shafting (see Fig. 81).
Quickly pour in the molten babbitt until the shafting is well
covered. When it has cooled, remove the centering and clamping
screws and withdraw the shafting. A greater amount of expansion
may have taken place in the casting than in the shafting and, upon
cooling, the bearing may be extremely tight. If the shaft cannot

be twisted free in the vise, hold a block of wood against its end and drive it through with a hammer. If it is still tight after being cleaned out, smear a mixture of cutting oil and rotten stone in the bearing, then push the shafting back in and work it around. This operation is known as lapping, and if it is not overdone a perfectly smooth turning action will result.

Now make up the declination axis tight in the heavy flange K, and slip one collar over it, up against the flange. Smear shaft and collar with the lampblack and oil, then approximately center it in tee E, and fix the second collar in position against the tee face. Some means of adjusting the two axes perpendicular to each other must now be found. The success of the final adjustments is dependent on the accuracy of this setting, so take plenty of time with it as it cannot be corrected after the babbitt has been poured, except by melting out the bearing. In order to adjust correctly the position of the declination axis, the polar axis must be temporarily screwed into the branch of the tee E, and the shaft then inserted in its bearing, already made. Rig up a makeshift pointer on the wye, contacting the declination axis at some point; then rotate the polar axis through 180° and see if the pointer again contacts the declination axis. By repeated trials, and adjustment of the supporting screws in the tee, a very accurate setting can be arrived at.

Carefully, now, so as not to disturb the setting of the declination axis, unscrew the polar axis from the tee. Support the declination assembly as in Fig. 81, and babbitt. Be very careful in pouring not to bring the level of the molten metal up too high in the branch opening, or it might not be possible to screw the polar axis all the way back in.

When cool, remove the shaft from the bearing, and use gasoline to flush out the lampblack, burned oil, and other foreign matter. If the bearing is tight, lap it in as already described. A bearing scraper is a useful tool for dressing up rough parts. Finally, cut two washers of leather or other friction material and insert them ahead of the clamping screws, to cushion the thrust of the latter against the shafting.

Screw the polar axis back into the tee, and pin it securely in its original position; then smear clean lubricating oil on both axes, return them into their bearings, and the assembly is completed.

Further Suggestions. The amateur who has access to a lathe will, no doubt, contrive to machine the faces of the tee and wye, and so effect smoother performance. With careful planning of the whole operation, they can be made square with the axes. Added refinements, which are left to the discretion of the worker, are a worm-and-gear slow motion on the polar axis, and either a rack-and-pinion or a multiple-thread focusing arrangement on the eyepiece adapter tube. Note the provision for making latitude correction in Fig. 72. It will pay to take a little time in accurately setting the four bolts (only three are visible in the diagram) in the concrete pier, as adjustment will be greatly simplified if they are exactly oriented. A method of locating true north will be found in Chapter XII.

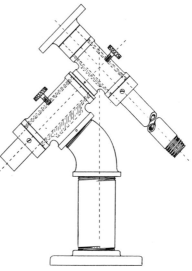

Fig. 73. A rigid pipe mounting, with babbitted bearings. Parts are listed in Table I, under Portable "B."

A Portable Mounting. Table I lists sizes and descriptions of parts for both permanent and portable mounts. A 45° elbow with a tee, used as shown in Fig. 71, rather than the wye of Fig. 72, is preferred for the portable mount, as it dispenses with the counterweight on the polar axis. The wye is more rigid, of course, and so is chosen for the fixed mounting. In thus dispensing with the counterweight, the radial thrust load on the polar bearing is increased and is in counter directions at the opposite ends. It may therefore be desirable to increase this bearing area by using the parts listed under "B" in the table. This is strongly recommended if an additional five or six pounds can be tolerated.

The gain is not only in an increased diameter of the polar-axis bearing, but also an increase in the length of both bearings, as a comparison of the proportions of the respective tees will show.

The additional length in the bearings is an important factor in gaining added rigidity. The mounting, detailed in Fig. 73, and weighing but 17 pounds, is the one shown in Plate VI (page 136).

As the difference in cost is little, brass fittings are to be preferred. Brass is easier to machine than iron, will not rust, and has a more pleasing appearance. Also, the inside surfaces of the brass fittings can be tinned, thereby providing a secure anchor for the babbitt. The standard brass flange is not sufficiently heavy to support the saddle and tube adequately, but there is a similar fitting known as a flange union, consisting of two heavy flanges (3½″ in diameter in the 1″ pipe size) which comes as a unit, and it may be possible to pick up an odd one. Machining steps will be described following the discussion of the tripod.

The Tripod. The combination of tripod and pillar (frontispiece) was devised in order that the telescope could be swung through the meridian without fouling the legs. The only other way in which this could be accomplished would be to extend the declination axis several inches beyond its bearing support, thereby increasing the bending moment.

No weakness was introduced by the pillar, which proved to be exceedingly rigid. In construction, it consists of a 9″ length of aluminum tubing, 4½″ outside diameter, with a ⅛″ wall. Tightly fitting caps at each end were built up of three thicknesses of ⅝″ plywood, glued together, then turned in the lathe and perforated to admit the pipe that is screwed to the mounting. This pipe passes all the way through the pillar and is made up tight at the bottom by means of a lock nut. The three 2″ blocks of maple to which the legs are attached were first glued to the lower cap, and included in the turning and boring operation.

For the f/8 telescope, the pillar construction may prove every bit as rigid if it consists simply of a 10″ length of pipe, extending from the 45° elbow, and a heavy flange screwed onto its lower end. The flange in turn can be bolted to the wooden tripod head, or the pipe can be screwed directly into the tripod head of metal shown in Fig. 76. If this form of construction is adopted, the tee that houses the polar axis should then be of the "bullhead" type shown in Fig. 73, with a 2″ branch opening.

The legs of the tripod are made of maple, although any hardwood will do. The slats are 0.4″ thick, 1½″ wide, and 21″ long. The oak spacers are 1⅛″ thick, the upper ones being the same width as the head blocks (2″), and the lower two tapered, narrowing down to 7/8″ at the bottom. The pegs or feet are made from

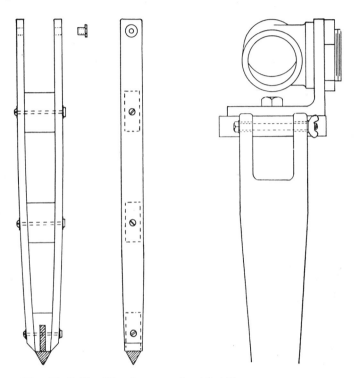

Fig. 74 (left). Plan for a wooden tripod leg.

Fig. 75 (right). How wooden tripod legs may be attached to the casting shown in Fig. 76. This diagram also shows the manner in which the telescope in Fig. 93 is hung on an angle plate bolted to the tripod head.

1″ bar stock, 3¼″ long. One half of this length is reduced to a neck with a diameter of ⅜″; the other half is tapered to a point. The neck is pushed into a ⅜″ hole drilled into the lower wooden spacer. When glued and bolted together, the leg assembly (Fig.

74) makes an extremely rigid truss member. A synthetic resin glue, such as weldwood or cascamite, mixed to a thick creamy consistency, is recommended. The overall height is 31″ from the floor to the top of the pillar with the legs spread at an angle of 60° to the floor.

The legs can also be fashioned from a solid board, as suggested in Fig. 75, but the board should be not less than 1¼″ thick and 3″ wide. As in the design in Fig. 74, brass bushings should be inserted into the holes to take up the wear against the joining bolts. See page 200 on how to tie tripod legs together.

A wooden pattern for the tripod head, similar to Fig. 76, is easily made, and a casting in iron, brass, or aluminum made up at a local foundry. The diameter is a matter of choice, the larger the better, but 5½″ is about the limit if the tube is not to tangle with the legs. A good thickness is about 1½″. Make the center hole about 1¾″ in diameter, and very slightly taper it and all the sides of the pattern in the same direction, so that the foundryman

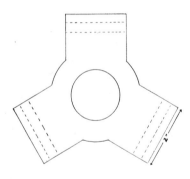

Fig. 76. Plan, laid out in a 5½″ circle, for a tripod head.

can remove it from the sand. To accommodate the 2″ pipe, the center hole should later be bored out to a diameter of 2.236″, and then threaded squarely with a 2″ pipe tap, sufficient threading being cut to admit all but about the last thread on the pipe when it is made up tight. The only other machine work required on the casting is the drilling of the three holes (5/16″ or ⅜″) for attaching the legs. The sides there should, of course, be filed until they are quite parallel.

Machining the Parts. The following suggestions of step-by-step operations for making a machined mounting similar to the one shown in the frontispiece might be thought to be useful only to those having access to machine-shop facilities. It has been suggested, however, that an amateur not equipped to do the work himself might farm it out to a local machine shop. The idea is a practical one, and while it is difficult to assay the cost of such a project, the figure is not likely to be much in excess of 20 dollars. After learning what is needed, the machinist can then either follow the steps outlined here or substitute other and better methods for achieving the desired results.

The parts used in this description are the same as those used in the mounting in the frontispiece and listed under "A" in Table I, but, as previously mentioned, substitution of the parts listed under "B" can be made if desired. On account of the small difference in diameter of pipe and fittings, the internal threads of the tees must be bored out to create sufficient space for the babbitt.

The external and internal threading on pipe and fittings is tapered, so that a tight joint can always be made up, but in so doing several threads always remain exposed. Furthermore, the pipe is never perfectly round, and may have scorings on it from the fitter's wrench or vise jaws. None of this annoys the steamfitter, but the telescope maker is going to utilize short lengths of this pipe (or long nipples, as they are called) for his polar and declination axes. To turn in bearings, these must have shafts that are truly round and smooth right up to the shoulders of the fittings to which they are joined.

The declination nipple screws into a heavy flange and the polar nipple into a tee, so the internal threads of both these fittings must be tapped deeper with a 1″ pipe tap, to permit further entrance of the pipe threads. This should be done a little at a time, and the nipple tried frequently, until when made up tight with a wrench only one or $1\frac{1}{2}$ threads remain exposed.

In using the wrench, apply it near the opposite end of the nipple or, better, to a spare fitting such as a coupling which has been temporarily screwed on there. This will avoid scoring or bending that part of the pipe to be encased in the bearing. Two pipe wrenches are convenient for working one against the other. Once the pipe has been rounded in the lathe, the wrench should never be applied directly to any part of it, for fear of deformation. Screw a spare coupling on the free end and apply the wrench to that.

When turning the nipples in the lathe, remove only a minimum amount of metal in order not to thin the walls any more than is absolutely necessary. Removal from the newly poured bearings will be made easy if, at the same time, the nipples are given a taper of about 0.001" for each inch of their length. Therefore, for those two operations, the tailstock center should be offset about 0.008".

1. Make up the declination axis nipple in the heavy flange. Hold it in the chuck, running true, and cut off any excess length that protrudes beyond the back face, recessing it to about 1/32" below the face of the flange. At the same time bore out the inside of the nipple to a depth of about 1/4", removing a minimum of metal.

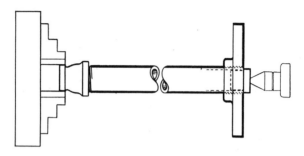

Fig. 77. **Operation 5** — Machining the declination axis.

2. Turn up from scrap stock a metal plug to fit into the above bored-out end. Return the first assembly to the chuck and center-drill into the plug for the tailstock center.

3. Make up the polar-axis nipple in the reducing tee. The end will be seen to project inside where it will be in the way of the babbitt, so after noting the length of the excess projection, remove the nipple and cut this excess off, after which the nipple is returned to the tee. Hold the nipple in the chuck, running true, and center-drill into the back of the tee for the tailstock center.

4. The tailstock center is now offset about 0.008" away from the worker; the diameter of the work will thus be tapered in the direction of the chuck end, where it should be least.

5. A spare coupling is screwed firmly onto the end of the declination assembly, and held in the independent chuck with the tailstock center supporting it at the other end (Fig. 77). Adjust the chuck to run true. Machine all parts indicated by the heavy lines in the diagram except the back face of the flange, which would be made conical instead of flat if done while the tailstock center is offset. To face this surface, remove the coupling and hold the

machined nipple in the chuck, running true. This will bring it reasonably square with the axis of the shaft. Reverse the assembly, holding the flange in the chuck, and turn off the tops of those threads on the other end of the nipple that had been covered by the coupling.

6. Screw the spare coupling on the free end of the polar-axis nipple, and hold it in the independent chuck, running true, with

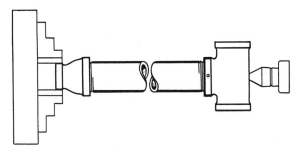

Fig. 78. Operation 6 — Machining the polar axis.

the tailstock center supporting the tee end (Fig. 78). Machine the surfaces indicated by the heavy lines. For further machining, it is necessary to remove the nipple from the tee, but first, to insure later return of the parts to the same exact positions, a small hole into which a pin or screw can later be inserted should be drilled through the two parts, as is indicated in the diagram. (A No. 44 drill, tap size for a 4-36 screw, will do.) After removing the coupling and unscrewing the nipple, reverse the assembly in the chuck and turn off the tops of the threads at the other end.

7. The faces of the reducing tee must now be machined, parallel to each other, and at right angles to the branch face. The small center hole in the back of the tee is enlarged to allow entry of a long screw which threads into a 3″ length of flat bar stock, which is then drawn firmly against the machined branch opening of the tee. Hold the tee in the chuck (Fig. 79) and adjust until a tool point just contacts the surface of the flat bar evenly, along its length, as the lathe carriage is moved along the ways. Thus insuring that the face of the flange will now be made quite square with the branch opening, remove the bar. The face and outer

shoulder of the tee can now be machined, and the internal threads at that end bored out. Reverse the tee in the chuck and repeat the operations at the other face, making both outside diameters equal. The faces should be parallel to each other within about 0.001".

8. Only one face of the second tee need be machined, but as it may be desired to put a collar on the lower end of the polar axis, it is best if both faces are machined parallel to each other. The internal threads must, of course, be bored out. This tee should first be experimentally made as tight as possible with close nipple and elbow, and the face that will be uppermost marked for identification, for example, by spot drilling into the shoulder with a small drill, or by filing a notch in the edge.

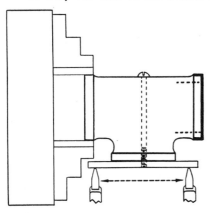

Fig. 79. Operation 7 — Facing off the tee, declination axis.

9. The backs of both tees are now drilled and tapped for the clamping screws L, Fig. 72. Tinning the inside of the tees to provide more secure anchorage for the babbitt is optional.

10. Four collars, preferably of aluminum because of its light weight, must be turned up either from solid stock or from castings, and bored out to suitable diameters to provide a neat fit on the shafting. Make the faces of these collars parallel and square with the boring. The outside diameters should be concentric with the borings, and of the same diameter as the shoulders of the tees. This will facilitate easy centering of the shafting when you are getting ready to babbitt. The width of the collar at the upper end of the polar axis should be such as to bring the intersection of the axes vertically over the center of the tripod (see Fig. 73). The collar at the upper end of the declination axis ought to be at least 1" wide. The width of the other collars is optional, and they should be provided with two set screws each, spaced 120° apart.

11. Three holes, 120° apart, near each end of both tees, should be drilled and tapped for No. 6-32 screws (Fig. 80) which support the pipe or shafting centrally for the babbitting operation. Also drill two small holes near each end of both tees, to allow the escape of air. The ends of the centering screws should be filed smooth to prevent their scoring the pipe. The ends of the ⅜″ clamping screws (*L*, Fig. 72), which can be made or purchased, are filed or faced off flat.

Fig. 80. How shafting is centered in the tees for babbitting.

12. The details of the babbitting operation have already been described in the making of the fixed mounting. About half the quantity of babbitt that was required there is needed for the portable

Fig. 81. Pouring the babbitt for the declination axis bearing.

mounting. The declination bearing is the first to be poured, and is assembled as shown in Fig. 81, after first coating the pipe with oil and lampblack. Be sure to insert the clamping screw. Be care-

ful not to fill too much with the babbitt. After cooling, remove
the centering and clamping screws, screw a coupling onto the end
of the nipple and drive it out with a few light blows of a hammer.
That end of the tee adjacent to the flange should first be marked
to identify it as the large end of the bearing.

13. Now screw the polar axis back into this tee until the
previously drilled hole (operation 6) is in alignment; there a pin
or screw should now be inserted. Coat the nipple with oil and
lampblack and make up the assembly as in Fig. 82, ready for
babbitting. The previously marked face (operation 8) of the tee
which is to house the polar bearing must be placed adjacent to the
collar that fits against the screw joint, and while it is held centered
the centering screws are brought up to just touch the polar axis,
enough to keep the assem-
blage from shifting. The
two bearings will thus be
brought at right angles
to each other with all
reasonable accuracy. The
second collar is brought
up flush against the face
of the tee, and held there
by means of its own set
screws. In preheating this

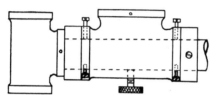

Fig. 82. Polar axis bearing ready
for babbitting.

assembly, keep the torch flame away from direct contact with the
babbitt of the declination bearing.

When the mounting is finally assembled, two wafers of leather,
vellumoid, or some friction material are cut out and inserted before
the clamping screw of each axis; these cushion the thrust of the
screws and prevent their scoring the shafting. After the pipe
nipples which constitute the declination and polar axes have been
trued up in the lathe, under no circumstances should they be held
in a vise, on account of the almost certain probability of their be-
coming deformed. Screw a fitting on the exposed threaded end,
and apply the force to that. Use two pipe wrenches, cautiously, in
removing the tightened fitting.

Chapter XI

ALUMINIZING AND CLEANING

Aluminizing. As mentioned in Chapter VI, it is best to defer the aluminizing until the mirror cell and other tube parts have been completed; then the mirror and diagonal can be treated at the same time. It seems almost superfluous to add that it is the polished and figured surfaces that are aluminized. The lustrous metallic coating, of the order of a quarter of a wave length of light in thickness, increases their reflectivity some 22 times.

Of course, the finished coating of aluminum should have no foggy areas or other blemishes. Pinholes in the aluminum film can be avoided, although it is quite common to find, upon holding the mirror up to a strong light, at least a sparse scattering of them rather evenly distributed over the surface. If they are not too numerous, the light loss thereby occasioned can be disregarded, but a profusion of pinholes, indicating that a poor vacuum was had in the process, should not be tolerated. It is desirable that the coating be perfectly opaque, but a loss in transmission of as much as one per cent of the light is tolerable.

Sometimes the aluminum has been unevenly deposited, with a considerable variation in its thickness in different areas, giving a spurious and detrimental figure to the mirror. On several occasions I have come across aluminized mirrors that bore no resemblance whatever to the original figures. Of these, one that was originally hyperbolic proved to be an oblate spheroid under test. It was again hyperboloidal after removing the aluminum, so any possibility of flexure was disproved. Another poorly aluminized mirror showed partial astigmatism, and another consisted of a nondescript assortment of shapes.

Accordingly, after your mirror is returned by the aluminizing laboratory, give it another knife-edge test, and for this it may be necessary to make a very tiny pinhole; otherwise the almost blinding reflection may render the shadows indistinguishable. A filter, however, such as colored cellophane, may permit testing with the old pinhole. Needless to say, the figure should faithfully correspond to the one that was originally given to the glass surface.

It may be of interest to record the condition of three aluminized mirrors figured by the author, which have been giving excellent performance. When mirror A was held about three feet in front of an unfrosted 100-watt lamp, a rather even sprinkling of tiny pinholes was observed. The filament was clearly and sharply outlined, steely-blue in color. As the mirror was moved about, in one or two places the filament became dim, suggesting that in those areas the film was more thickly deposited. When the mirror was held up to the sun, there was considerable irradiation in yellow and blue light, and because of its brightness, the outline of the sun could not be seen. No attempt was made to measure the amount of light transmitted, but best estimates placed it at under one per cent. Mirror B showed but a thin scattering of pinholes, and was apparently opaque to the electric lamp. The sun was visible through it, quite dim, but hard in outline and whitish in hue. There was little variation in its appearance through different parts of the film, although in one region it became nearly obscured. Except for about half a dozen pinholes, mirror C was absolutely opaque to all visible radiation; not even the shadow of a solid object passed across the path of the sun's rays was observable. On the Foucault stand, none of these mirrors revealed any noticeable departure from its original figure, nor could any criticism be made of the appearance of the extra-focal diffraction rings of a star, despite the fact that, in the cases of A and B, variations in the thicknesses of their respective films were demonstrated.

Because of its relatively small area, there is little likelihood of the figure of the diagonal being affected to any injurious degree, although the figures on larger optical flats have sometimes been altered by aluminizing. In rechecking by the interference test (described in Chapter VII), there is great danger of the aluminized surface being scratched unless suitable precautions are taken in bringing the surfaces together. Any dust particles will be conspicuously visible on the diagonal's surface, and may be blown off with the breath. Or if they still adhere, the smooth skin on the under side of the forearm, after being washed and thoroughly dried, makes an effective "wiper" if passed lightly across the face of the diagonal. The test piece, similarly freed of dust and other foreign matter, is then carefully placed on top of the diagonal, and lightly but firmly pressed down, slowly excluding the air until contact is made. It should not be allowed to slide about in this process, as scratches will almost surely develop.

Neon light is practically useless for this diagonal test, for the almost overpowering reflection from the aluminized surface makes the interference bands so faintly discernible that one may falsely conclude that some contamination prohibits their appearance. A more monochromatic light source is certainly desirable for testing silvered or aluminized plane surfaces.

Magnesium fluoride coatings were mentioned in the first chapter; their purpose is to reduce reflection losses at the surfaces of lenses and prisms. If a prism is used for the secondary reflection, its square faces should be fluoride coated, of course. Or the diagonal flat, if used, as well as the primary mirror in any case, should be similarly treated to prolong indefinitely the maximum efficiency of the aluminum film. The fluoride coating provides a hard, durable surface that withstands abrasion well without impairing the optical qualities of the surfaces.

A silver coating is not as durable as aluminum, but it may be put on by the amateur himself. Instructions for silvering by the Brashear and other processes may be found in optics, telescope making, and physics laboratory texts.

Packing for Shipment. Protection of the surface of the mirror from abrasion during transportation to the laboratory is imperative and easily accomplished. Place two or three thicknesses of absorbent tissue over the concave face of the mirror, and on top of that place a clean thin board, such as plywood, about 6½″ square. On a second similar board spread some absorbent cotton, and lay the mirror on top of that. Cut out some strips of corrugated paper, less than an inch wide, and wrap them around the mirror's edge. Then bind the whole together with adhesive tape, holding the mirror immovable in a compact package. This is then placed in a container of metal, wood, or corrugated board, surrounded by ample cushioning material. If the diagonal is to be included, cut the center out of a piece of corrugated board and place it on top of the package containing the mirror; wrap the diagonal in cotton and put it in the center hole, place another sheet of corrugated on top, tie all together with string and place in the container. As a precautionary measure, an addressed tag should be attached to the inner parcel.

Care of the Optical Parts. Because of the nature of the reflecting telescope, its owner is powerless to do anything in the way of protecting the mirrors during exposure, with the result that dust, dirt, and the elements get in their licks unopposed. But a reasonable amount of protection can be given to the optical parts when they are not in use. Of course, a perfectly designed cell is one that permits the mirror to be quickly removed from the telescope and replaced again when needed, without altering the direction of its axis. In this way the mirror can be given complete protection between observations. The diagonal can be equally well protected by designing a dust-tight cap to fit over it. Such arrangements have been devised by a number of amateurs and have proved to be most practical. But since the tube can be brought indoors (not to be stored in a damp cellar), the mirrors may remain in their places, and at the same time can be protected with tight covers over each end of the tube. Cap-like refrigerator dish covers, made of rubberized fabric with hemmed-in elastic bands, are serviceable for this purpose. The eyepiece opening should be plugged up or similarly covered.

Although it is not a regular occurrence, dewing of the mirrors may take place whenever the temperature falls below the dew point of the night, and ordinarily there is no way of preventing it. Dewing will first be noticed on the eyepiece lenses or on the secondary mirror, at which time it will be wise to seal up the mirror end of the tube with a tightly fitting insulated cover. This will probably forestall condensation on the mirror long enough to enable the observer to conclude the evening's program. Eyepiece lenses, and the surfaces of a prism (if employed for the secondary reflection) can be safely wiped dry with a soft cloth or absorbent tissue, but it is not practical to attempt this in the dark on an aluminized diagonal. Pieces of blotting paper, cut to the size and shape of the diagonal, may be kept on hand for such an emergency, and a piece can be pressed lightly against the dewed surface. If any dirt is present no scratches will result, and the dirt and any lint left by the blotter can be removed at a more favorable time.

It is not necessary to bathe the mirrors each Saturday night. At long intervals, most of the thin layer of dust that will slowly accumulate despite the best of care can be blown off. Washing should be resorted to as infrequently as possible, as each washing further thins the metallic film. A mild face soap and distilled water or freshly trapped rain water should be used for the bath. Immerse the mirror in a soapy lukewarm solution and allow it to remain there for a time, so the surface dirt can be dissolved. Then, with the mirror still submerged, swab its surface lightly with sterile absorbent cotton, renewing the latter if necessary. A gray surface film resulting from chemical action of the elements may be present on the mirror, and this can be pretty well removed by rubbing a little more briskly with the cotton *after* all of the coarse dirt particles have been washed off. Finally, flush all the soapy water away in clear distilled water, and spread a clean handkerchief over the surface to dry it.

Occasionally, you will hear of a camel's-hair brush being recommended for cleaning the mirror. Now, the manufacturing optician often finds it necessary, in such operations as testing, cementing of lenses, or sealing up of lenses or prisms, to remove every last particle of dust, and the camel's-hair brush is an admirable tool for the purpose. But the reflecting telescope owner is not interested in isolated dust particles; he waits until there is a sufficient accumulation and then a bath is the only safe method of removal. Furthermore, unless the brush is drawn across the mirror very gently, and held at a low angle in so doing, the ends of the bristles are almost certain to leave sleeks (fine hairline scratches) in the soft aluminum film.

Eyepiece lenses, when soiled, should also be removed from their cells and given a bath. In replacing them, avoid leaving finger marks on the surfaces, as the perspiration oils contain an acid that has a corrosive action on the glass.

Chapter XII

SETTING CIRCLES–EQUATORIAL ADJUSTMENT

IN ORDER THAT THE MAXIMUM BENEFIT may be derived from its use, the permanently mounted telescope should be in accurate equatorial adjustment and equipped with setting circles. When the circles are finally adjusted and fixed, the telescope can then be set to the known declination and hour angle of a celestial object, and it will be found to be in the field of view. The hour angle is the difference in time between the hour circle of the object and the observer's meridian; in other words, it is the difference between the right ascensions of the object and the observer's meridian. The right ascension of the meridian is always equal to the local sidereal time. An example of how to convert local civil time (LCT) to local sidereal time (LST) will follow shortly.

Making the Circles.   A ready way of procuring the setting circles is to purchase two cardboard protractors 6″ in diameter. These come accurately engraved on Bristol board, with each quadrant marked from 0° to 90°, and graduated to half degrees, which is as much accuracy as the amateur astronomer will generally have use for. The protractors can be cut out and glued to plywood disks having accurately bored-out centers, varnished, and slipped over any convenient parts of the mounting. Pointers or indicators must likewise be attached. It is best to have the pointers attached to a stationary part of the mounting, while the circles rotate with the axes. The method of numbering on the protractors is ideally suited for the declination circle, but an alteration will have to be made for the hour circle. Change each quadrant so that, instead of containing nine major divisions of 10° each, it will be divided into six divisions reading from 0 hours to 6 hours, each 1° division then being equal to four minutes of time. The hours will be read in either direction starting at 0 hours.

The cardboard circles will render good service for a time if protected from the weather, but the construction of a more permanent set should be attempted while you are learning how to use the first ones. Experience with the cardboard circles may prove that half-degree graduations are not necessary.

Lacking the means of mechanically making precise graduations on metal rings or disks, you may transfer the divisions, as accurately as patience will permit, from a protractor or other suitably graduated device onto durable plastic disks. An excellent engraving tool can be made from an old hack-saw blade by grinding a hook-shaped cutter into it close to one end, so that in being drawn across a surface, bearing against a straightedge (just as in using a pencil), a clean groove will be cut. The hook edge is ground on the corner of a grinding wheel, and that part of the blade is also thinned down against the wheel until it is 0.008″ or 0.010″ thick, so that grooves of similar width will be cut. If this is carefully done, with the lines cut a few thousandths of an inch wide and deep, and filled with a contrasting color of paint or sealing wax, the circles will be both neat and serviceable. Sealing wax can be dissolved in wood alcohol, rubbed into the grooves with the fingers, and the surplus scraped off of the surface with the straight edge of a stiff piece of paper, such as a business card.

Adjustment of the Tube. It is assumed that the polar and declination axes of the mounting are perpendicular to each other. The mirror's axis, presumably coincident with the axis of the tube, must now be brought perpendicular to the declination axis. Place both declination axis and tube in a horizontal plane, and clamp the polar axis. Position the tube by rotating it in its saddle so that the eyepiece adapter tube is parallel to the ground. Plant two stakes in the ground exactly in line with the center of the tube and on opposite sides, each about 200 feet distant. If this much room is not available, use of a longer adapter tube to hold the eyepiece may enable you to bring the stakes into focus even though they are closer. The stakes should be plumb in the line of sight. Look into the low-power eyepiece, and the first stake should be seen in the field of view. Swing the tube on the declination axis through 180°, and the other stake should be seen in the field.

By appropriate shimming at either end of the saddle, both stakes must be brought exactly to the center of the field. A high-power eyepiece should be used for a final determination. Watch out for warping of the wooden saddle after these adjustments are made.

If sufficient ground space is not available for the above test, defer alignment of the mirrors until the mounting is built, and use the empty tube, fitted with cardboard disks at each end, one disk having a 1/16" centered peephole, and the other a ½" central aperture. This will provide a field of view of a little more than ½°. Plant the stakes as far off as possible, or better still, suspend plumb lines. Sighting through the peephole at first one and then the other plumb line, bring them both to the center of the ½" aperture by shimming as described.

Adjustment of the Polar Axis. At the present time, Polaris is 51 minutes of arc from the celestial pole and describes a diurnal circle of that radius about it. Twice in each 24 hours, therefore, at upper and lower *culmination,* the star is exactly north. And twice daily, at eastern and western *elongation,* it is at exactly the same altitude as the celestial pole. We shall first aim the polar axis true north, and later make the adjustment for latitude (altitude of the pole). Let us suppose that we are ready to begin the adjustments on the evening of May 17, 1973. The telescope is located in latitude 40° 52'.0 north, and longitude 73° 46'.2 (4ʰ 55ᵐ.1) west. Our watch is set by shortwave time signal to Eastern standard time (EST), one hour slower than daylight saving time, and five hours slower than Greenwich civil time (GCT), which is also called Universal time (UT). (Other time zone corrections in the United States are: Central. standard, six hours slower than GCT; Mountain standard, seven hours; Pacific standard, eight hours; Alaska-Hawaii standard, ten hours.)

In planning observations of celestial bodies, it is very helpful to have the American Ephemeris and Nautical Almanac (price about $6.25), published annually and for sale by the Superintendent of Documents, U. S. Government Printing Office, Washington, D. C. 20402. The arrangement of this publication is the same each year, so it will be easy for the reader to apply to any date the calculations in this discussion, which have been based on the 1973 *Ephemeris,* and to make appropriate substitutions.

Another observing aid quite useful to the amateur is the annual *Graphic Time Table of the Heavens*, a 40″ by 27″ wall chart available from the Maryland Academy of Sciences, 119 S. Howard St., Baltimore, Md. 21201, for about $2.00. On a reduced scale, this chart is also published in each January issue of the magazine *Sky and Telescope*. While the *Ephemeris* is the standard astronomer's tool, and examples of its use are provided later, all of the adjustments of the equatorial mounting can be performed about as well with the *Graphic Time Table*, since accuracy to a few minutes is sufficient. Polaris' position in the sky changes so slowly that critical timing is unnecessary.

To make the first adjustment of the polar axis, examine the line on the *Graphic Time Table* representing the night of May 17-18, 1973. At about 10:20 p.m. local time, a slanting line marked "Polaris — Lower Culmination" crosses the date line. This is the local mean time when Polaris crosses our local meridian below the pole itself. But our watch is set for the 75th meridian, on which Eastern standard time is based, and since we are 1° 13′.8 east of that meridian (4^m $55^s.2$ in time units), the watch is running about five minutes behind local time. In other words, Polaris reaches lower culmination at about 10:15 p.m. watch time (Eastern standard) on the evening of May 17th. (The choice of New York City's longitude produces a similarity of numbers that should not be confused with each other. The 4^h $55^m.1$ represents our distance *west of Greenwich*, and 4^m $55^s.2$ our distance *east of the 75th meridian*.)

At intervals of about six hours (5^h 59^m watch time), further Polaris events will follow in this order: eastern elongation, upper culmination, western elongation, and lower culmination again. The intervals between events are not strictly equal, but no significant error is introduced for our purposes by considering them equal. It is obvious, however, that only one other event can be observed by us on the night of May 17-18, an eastern elongation of Polaris at 4:14 a.m. EST.

So at about 9:45 p.m. on the 17th, set the declination axis in a horizontal position and clamp the polar axis. Point the telescope northward and sweep in a vertical plane, in order to pick up the North Star, which must be brought to the center of the field by rotating the wye fitting east or west on the pier (Fig. 83). Try to accomplish this by 10:15 p.m. as nearly as possible. Then set the hour circle so that the pointer is on zero, and note the reading on the declination circle. Release the polar axis and turn it through 180° by the hour circle, so that the declination axis is again horizontal, with the telescope on the other side of the pier. Clamp the polar axis, and bring Polaris to the center of the field by sweeping

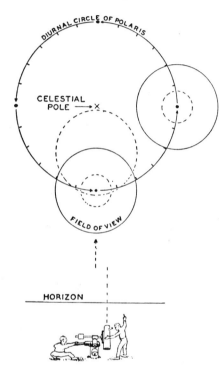

Fig. 83. Adjusting the telescope in azimuth, which, in the sketch, is being done at the time of lower culmination of Polaris. The distance of Polaris from the celestial pole is greatly exaggerated in the drawing. The diurnal circle is marked off in hours, and the small gaps at the positions of elongation and culmination represent the apparent motion of Polaris in 20-minute intervals. "Field of view" is the real field that will be taken in by a low-power eyepiece. The smaller dotted circles represent the comparative field of view of a ½" eyepiece.

in a vertical plane as before. Again note the reading on the declination circle. It is not likely that both of the readings will be the same. Now move the tube so that the declination pointer just splits the difference between the two readings, clamp it, and bring Polaris to the center of the field by tilting the mount up or down by means of the north-south lock nuts on the pier. For better accuracy, the adjustments should be completed with a high-power eyepiece.

Because of Polaris' nearness to the pole, its diurnal motion across the line of sight is slow (see caption in Fig. 83), and for completion of these preliminary adjustments a period of 20 minutes —10 minutes before and 10 minutes after the moments of culmination and elongation—may be allowed. This will not affect the final results, since the observations will have to be repeated anyway.

The polar axis now points to Polaris, which, within the interval just mentioned, is substantially due north and 51′ below the pole. A partial correction for altitude may be made by raising and lowering, respectively, the north and south lock nuts, tilting the polar axis upwards, so that Polaris is brought to the upper edge of the *inverted* field (large dotted circle, Fig. 83, inverted).

Polaris will reach eastern elongation at 4:14 a.m., which is less than half an hour before sunrise at this time of year, but the 2nd-magnitude star should be easily observed in the twilight sky. No harm will be done by completing the work 10 or 15 minutes ahead of time, since Polaris' altitude will be changing by only a minute of arc for every five minutes of time (see Table II beginning on page 512 of the 1973 *Ephemeris*).

So begin working again about 3:30 a.m., setting the telescope as in Fig. 72, with the declination axis in a vertical plane and with the tube above the pier. Clamp the polar axis, and bring Polaris to the center of the field by continuing the adjustment of the north-south lock nuts. Finish up as nearly as possible to the 4:14 position of Polaris, and set the declination circle to read 89° 09′.

On the following night, May 18th, we shall want to check the adjustments on another star, preferably one near the celestial equator. The choice of a star can be made by referring to the meridian transit lines for stars in the *Graphic Time Table*, or by paging through the tables on the mean places of stars in the *Ephemeris*. The simplest way is just to observe the evening sky at this season, whence it is seen that the 1st-magnitude star Spica, Alpha Virginis, is a good choice, and will transit about 9:30 in the evening, an hour before which time it is sufficiently dark to begin.

Choosing 8:30 p.m., May 18th, as our time for starting to work, we compute the meridian angle (negative hour angle) of Spica for that time. This will give its distance east of the meridian and enable us to predict easily the time it will cross the meridian. The methods employed are several, and well known to navigators.

If the *Graphic Time Table* is used, the procedure is simply to find the right ascension of Spica in a star catalog and to plot its transit on the chart. The line should be made parallel to the lines of other stars (parallel rulers will help) and crossing the midnight line where the sidereal time equals the right ascension. The local time of transit for May 18-19 can then be read.

The telescope maker, however, will do well to use the method whereby the local sidereal time is first computed (unless a timepiece running on sidereal time is available), and then the star's right ascension is subtracted from the sidereal time to give the hour angle. If the answer is negative, the star is east of the meridian an amount called the meridian angle, as already mentioned.

The Sidereal Time and Mean Places sections in the *Ephemeris* are used. Spica's position is given on page 338 as R.A. 13^h 23^m 46^s.0, Dec. −11° 01′ 15″.

Sidereal time 0^h UT, May 18	15^h	42^m	08^s.8
20^h 30^m May 18	20	30	
Correction for 20^h 30^m		+03	22.1
Longitude correction, 4^h 55^m.1			+48.5
Sidereal time 8:30 p.m. LMT	12^h	16^m	19.4
Correction standard time (sidereal interval)		04	56.0
Sidereal time 8:30 p.m. EST	12^h	21^m	15.4

The difference between this sidereal time and Spica's tabulated right ascension gives us an hour angle of 1^h 02^m 30^s.6 east, so that Spica should transit 1 hour and 3 minutes after 8:30, or 9:33 p.m.

Now aim the telescope so that the pointers read 11° south on the declination circle, and 1^h 03^m east of the meridian on the hour circle, and see if Spica is in the field at 8:30. Rotate the mount east or west on the wye to bring the star to a middle division of the field of view, while following it to culmination. At the moment of transit, 9:33 p.m., the hour circle is set to read 0 hours, and the mount is tilted up or down to bring the star to the center of the field. The declination circle is then set to 11° south.

Repeat the observations of Polaris on another evening, and again on a third star, one that is two or three hours east of the meridian. With patience, and if the mounting has been well built, it will be possible to bring the adjustments within the accuracy of the graduated circles.

Where possible, avoid stars having an altitude of less than

10°, as errors due to atmospheric refraction then become considerable and must be taken into account. Therefore, amateurs residing in low latitudes should apply a correction for refraction when making adjustments to Polaris. For example, at a latitude of only 10° north, Polaris appears 5′ or 6′ higher in the sky than it really is, and at 5° north the correction is 10′.

Sidereal Time from the Stars. When the adjustments have finally been completed, sidereal time can be found without computation by observing the transit of any celestial object, the right ascension of which is known. Its right ascension is, at the moment of transit, equal to the local sidereal time. Or, at any time, center the object in the field of view, and note the hour angle as read by the indicator. Adding this to or subtracting it from the right ascension of the object, depending on whether it is west or east of the meridian, will give the local sidereal time.

It will be found more convenient, however, to set a spare timepiece, and to keep it running on sidereal time on nights of observation. The slight difference in the rate of an ordinary watch will not matter during one night's use. Observatory clocks run at the sidereal rate, making computations unnecessary. Amateurs may purchase sidereal watches and clocks from a few American manufacturers.

The *Graphic Time Table* and the *Ephemeris* explain the method of finding sidereal time, but for a more complete study of the various kinds of time, the reader is advised to consult a standard astronomy or navigation textbook.

Locating True North by the Sun. To find the watch time of meridian transit of the sun, in order to locate true north by day, we must know the exact longitude. This can be found with sufficient accuracy from the topographic maps of the U. S. Geological Survey. Because of the *equation of time,* a difference between the clock and the sun caused by the ellipticity and inclination of the earth's orbit (as referred to the celestial equator), the time of "ephemeris transit" must be looked up in the *Ephemeris* for the date in question. This time, after adjustment for the observer's longitude, is what his watch will say when the sun transits.

For example, to find the watch time of transit of the sun on May 18, 1973, at a place in longitude 73° 46'.2 west, turn to page 27 of the *Ephemeris*. For all practical purposes, it can be assumed that the time of "ephemeris transit" of the sun equals the Universal time of its transit across the Greenwich meridian (an error of much less than a second is introduced). The first step is to compare the listings for May 18 and May 19, and interpolate for the observer's longitude:

UT of Greenwich transit, May 18	11^h	56^m	$21^s.48$
UT of Greenwich transit, May 19	11	56	23.95

The difference is $2^s.47$, which is multiplied by the fraction $73°.8/360°$ to get $0^s.51$. Adding this to the May 18 time and rounding to the nearest second gives:

LMT of local transit, May 18	11^h	56^m	22^s
Correction to standard time		04	55
Watch time of local transit	11	51	27

Set up a plumb line over the pier, and mark the direction of its shadow at this moment of the sun's transit. This will be a true north-south line. The watch will, of course, have been accurately set by radio time signal.

Chapter XIII

OPTICAL PRINCIPLES — ATMOSPHERE — MAGNITUDES

We found in chapter ix how magnification in the telescope is brought about. But this is not the only function of a telescope when used astronomically; depending on the uses to which an instrument is put, resolving power and light-gathering power may be of considerably more importance. A knowledge of some of the optical principles associated with these functions will enable the amateur to operate his own telescope to best advantage. The discussion here is necessarily brief, and may be supplemented by reference to various books listed in the bibliography.

Magnifying power has already been discussed, but in what follows it will become evident that light-gathering power and resolving power often limit the magnification which may be usefully employed. Magnifying power is otherwise independent, however, of the diameter of the objective lens or primary mirror, whereas resolving power improves as the aperture is increased and light grasp increases with the square of the aperture. The condition of the atmosphere through which we must study the sky is often the most serious factor in limiting resolution and magnification.

For visual work, the eye itself functions as the last optical system through which the light passes, the lens of the eye focusing the incoming light on the retina, which is situated at the back of the eyeball. In bright daylight, the iris of the eye contracts to a diameter of about 2 mm., and in darkness expands to 8 mm. or more; the iris thus acts as a stop in regulating the amount of light that is allowed to enter. The retina is composed of thousands of microscopic nerve endings, known as cones and rods. The cones are located near the central part of the retina, their greatest concentration being in a spot called the fovea. Acute vision is centered on the fovea, and is confined to a field of less than 1°. It is difficult to realize that this field is so small because the eye is unceasingly roving about, picking up and discarding one object after another. Extra-foveal vision, in which the rods are chiefly engaged, may extend up to 130° or more. The rods are much more sensitive to low levels of illumination than are the cones, but as much as half an hour in complete darkness may be required to adapt them totally for night vision. Because of the location of the rods on the retina, faint objects are most readily found with averted vision, or "out of the corner of the eye." If the object is sufficiently faint, any attempt to focus its image onto the fovea for careful scrutiny will actually result in its disappearance.

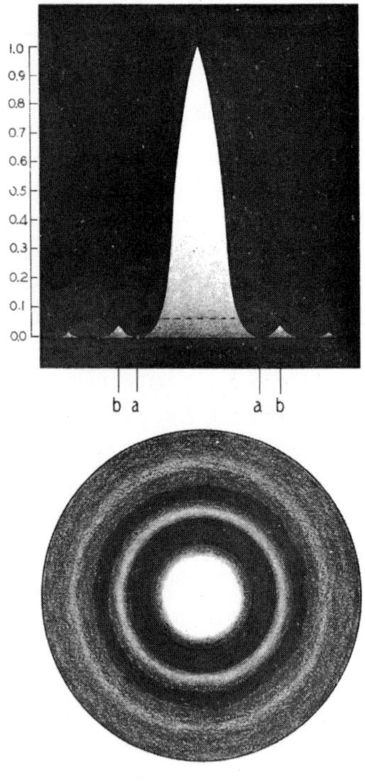

Fig. 84. Top: An imaginary solid showing the maxima and minima of the diffraction pattern. Bottom: An enlarged drawing of the diffraction image of a brilliantly illuminated artificial star, as observed under high magnification at the focus of a mirror stopped down to a ratio of f/24. With the brighter stars, the central disk will be relatively larger than shown, and somewhat smaller with faint stars. Practically, the diameter of the visible disk of any imaged point may be regarded as approximately one third that of the first bright ring or, in angular measure, 4.5 seconds of arc/aperture.

The stimulus caused by the light which passes into the eye is imparted to the cones or rods and interpreted by the brain as a picture of the object. As these nerve fibers are not in contact with one another, but are slightly separated, the picture that is seen actually consists of a multitude of adjacent areas. The size and central separation of these areas depends, in general, on the size and distribution of the fibers, much in the same way that resolution of detail in a photograph is limited by the graininess of the film.

In the focal plane of the telescope the image formed by the lens or mirror also consists of a multitude of nuclei, their angular sizes determined by the diameter of the mirror and the length of the light wave under consideration. The value of aperture becomes apparent when one realizes that with increasing diameter of the objective these dots (diffraction disks) become smaller and more light is poured into each of them by the lens or mirror. Thus, the larger the telescope, the greater the amount of detail that can be seen, other factors permitting. In discussing this subject, it is usual to consider a point source of light, such as a star.

The Diffraction Image of a Star. So tremendous is the distance of the stars that even such giants as Mira and Betelgeuse, whose diameters correspond approximately to that of the orbit of Mars, subtend such tiny angles (roughly 0.05 seconds of arc or less) as to be regarded as mere points. And on account of the wave nature of light, the image of a point source formed by any optical system is not itself a point, but a diffraction disk of finite size (Fig. 84), known as the Airy disk, after Sir George Airy, who in 1834 calculated the size and distribution of light in the diffraction pattern.

With a perfect optical system, about 84 per cent of the light is contained in the diffraction disk, with the other 16 per cent distributed in surrounding rings, one or two of which may be visible, depending on the brightness of the star and the stability of the atmosphere. With a mirror that is figured to within $\frac{1}{8}$ wave of a paraboloid, thus performing within the Rayleigh limit (see page 77), about 68 per cent of the light is contained in the central disk; the remainder is distributed in the rings. For a quarter-wave departure from the paraboloid, doubling the Rayleigh tolerance, only about 40 per cent of the light is in the central disk, although there is no noticeable increase in its size. The lost light is deposited through the diffraction pattern, fogging up the background, so there is a loss of contrast and definition except for very brilliant objects. With further departure of the mirror's surface from a paraboloid, there is apparent enlargement and fuzziness of the Airy disk, and definition and resolving power may become seriously impaired.

The appearance of a perfect star image as viewed in the telescope is similar to what might be seen in viewing from above a cone-shaped solid like that in Fig. 84. (The effect is simulated by inclining the page at a steep angle to the eye and squinting down the cone.)   The illumination is most intense at the tip, gradually diminishing down the sides of the cone, coming to a minimum at the base *aa*, then rising to a second maximum, *bb*, to form the first bright ring. The linear diameter in the focal plane of the first minimum (*aa*) or of the theoretical disk containing about 84 per cent of the total light, is:

$$d = 1.22 \ y \ F/r,$$

where *y* is the wave length of light (0.000022″), *F* is the focal length, and *r* is the radius of the mirror.

It is thus seen that the linear size of the theoretical disk varies with the focal ratio. But since the eye, in the presence of the intense illumination at the center, is not appreciative of the fainter light approaching the minimum, the visible diameter of the central disk is somewhat less. It is equal to the diameter of the cone in the plane of farthest descent of vision, as indicated by the dotted line in Fig. 84, and this will depend on the brightness of the star observed. In the case of a very bright star, the greater size of the visible disk and irradiation on the retina caused by a brilliant light source together make such a star appear to be considerably larger than a faint one. The image of a very faint star appears considerably smaller, the feeble illumination giving the impression that the image is a mere point.

The linear diameter in the focal plane of the first bright ring (second maximum) is $d = 1.62 \ y \ F/r$.

Resolving Power.   The *angular* diameter of the first minimum, *aa*, in the focal plane is equal to 11/*A* seconds of arc, where *A* is the aperture of the telescope in inches. It is thus an inverse function of the diameter of the mirror, and leads to a determination of the ability of the telescope to resolve detail. Theoretically, two adjacent points can be seen as separate when the center of the diffraction disk of one falls on the first minimum of the other, that is, when the separation of the centers of the disks is not less than the semidiameter of the first minimum, or 5.5/*A*. For a 6-inch telescope, this is about 0.92 seconds of arc. But, as has been shown by W. R. Dawes, a noted 19th-century observer of double stars, the least angular separation of two stars that can be seen as double is somewhat less (Fig. 85), and is given as Dawes' limit, a in the following formula:

$$a = 4.56/A.$$

This establishes 0.76 seconds of arc as the minimum separable angle, for points equally bright, in a 6-inch telescope.

With double stars of this order of separation, a considerable amount of magnification must be employed to enlarge the angle sufficiently for the eye to define the separate disks, presumably to the minimum angle that can just be resolved by the eye. The resolving power of the eye, or visual acuity, is the angle at the eye of two adjacent points of equal brightness that can just be resolved, and below which no distinguishable difference in dimension is apparent. The minimum angle is taken to be about 1 minute of arc with a pupil diameter of 4 mm., although acuity is fairly constant for pupillary openings between 2 mm. and 5 mm. Below 2 mm. there is a rapid decline due to the increasing size of the diffraction pattern on the retina; above 5 mm. chromatic and spherical aberration of the eye effect a decrease in acuity.

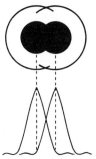

Fig. 85. Minimum resolvable separation of diffraction images, based on Dawes' law, of two equally and moderately bright stars.

For most eyes, 1 minute of arc also seems to be the limit in comparing objects of different widths. Objects of a much smaller angular size may be easily seen, however. The *limit of visibility* is a matter of contrast, as in the case of a fine wire stretched across a background of cloudy sky, which wire may still be visible when its angular size is only 1/50 of a minute. For objects placed against a dark background, as in the case of the stars, there is almost no limit.

But although the theoretical resolving power of the eye is usually taken to be 1 minute of arc, as stated above, experiments by the author with a group of students indicate that the eye does not actually perform down to this limit with separate luminous points. Pinhole perforations of the order of 0.0003″ in diameter were made in pairs in the painted surfaces of glass slides, so that at 10", the distance of best vision, their angular separations ranged from 1 to 4 minutes. Conditions were such as to provide a pupil opening of about 4 mm. Where there was more than a slight difference in pinhole diameter or brightness, resolution was limited to from 2½ to 4 minutes of arc, but with equal pinholes, and moderate illumination, several students were able to resolve a 2-minute separation. Only one student succeeded in resolving an equal pair of 1½-minute separation. As is theoretically to be expected, resolution deteriorated with brilliant illumination of the pinholes, and with unequal components.

It is rather well known that only a very keen eye can separate the wide components of Epsilon Lyrae, of magnitudes about 4 and 5, and 3½ minutes of arc apart. The inequality in brightness of the components is partially responsible for the difficulty of resolution, and the lack of stability in the atmosphere is also a factor.

A more liberal value than the 1-minute minimum must therefore be taken for use with the telescope, and in practice 4 minutes of arc has been found to be about correct. For a 6-inch telescope, then, to reveal as double two nearly equal stars separated by 0.76 seconds of arc, a magnification of 316, a little better than 50 per inch of aperture, must be employed.

With the f/8, so high a power is not generally apt to be used, due to the difficulty of acquiring an eyepiece of the requisite focal length. Occasionally, however, one is able to pick up an eyepiece with a focal length of 1/6" or even ⅛". But if the atmosphere is tranquil enough and the mirror is well figured, a pretty close approach to Dawes' limit can be expected with a power of about 200.

To resolve detail in an extended object such as a planet, the least magnification is one that will make the diffraction disk just visible to the eye; in other words, magnification that will enlarge the disk to an apparent angular diameter of 1 minute. This is found to be about 78x for a 6-inch mirror, or 13 per inch of aperture for any telescope. But the effort to detect detail at the limit of resolution imposes a severe strain on the eye, so some additional magnification must be used. For lunar, planetary, and terrestrial observations, probably up to four times this amount may be usefully applied, but further magnification will actually result in a loss of definition, because of the visibility of the separate diffraction patterns. A vivid demonstration of this can be made by trying the diaphragm shown in Fig. 87 on the planet Jupiter, for example, on a quiet night, first with a ½″ eyepiece, and then with a ¼″, and comparing the results. Except for the separation of close double stars, no benefit is to be derived from magnifications in excess of about 50 per inch of aperture.

Diffraction from Obstructions. The effect of obstruction by the vanes or spider legs supporting the secondary is to interrupt the continuity of the diffraction rings, the scattered light brightening the arcs adjacent to the interruptions. The visible effect is the ray-like radiation fanning out from a bright star image. Each vane causes a complementary ray to be formed on the opposite side, so that if the diagonal is supported by three vanes, six rays will be seen. With a four-legged spider, the complementaries coincide with the original diffraction, and only four rays result.[1] Increasing the thickness of the vanes increases the amount of diffracted light. In photography, the rays are often useful, as they make it possible to locate exactly the center of a star image which, due to long exposure, may be of enormous size on the plate.

The effect of the central obstruction is to reduce the size of the Airy disk, and to scatter more light throughout the pattern, at the expense of the intensity in the central disk; additional diffraction rings may thereby be made visible around a star image. When a 3″ stop was tried on Vega, the central disk shrank almost to a point, and numerous rings appeared. It was quite evident that there was a substantial increase in the amount of scattered light. Diffraction effects are more pronounced when a square stop, such as one in the shape of the obstruction from a rectangular diagonal, is used, although diffraction is kept at a minimum when the corners are made to coincide with the vanes, as in Fig. 63. In any other position, the corners create four additional rays radiating from the bright star image. The practice sometimes followed of reducing the area of the obstruction by cutting off the corners of the rectangular diagonal is not a wise one, as additional diffraction is introduced by each additional corner.[2]

---

[1] A discussion on how effectively to eliminate the rays caused by diffraction, by suitably curving the vanes, is contained in *Scientific American*, June, 1945.

[2] See Appendix VII, Diffraction Images, *Telescopes and Accessories*, by Dimitroff and Baker.

It has been found that the scattering of light, which impairs contrast, definition, and resolving power, becomes conspicuous when the area of the central obstruction exceeds six per cent of that of the mirror. Thus, in order to keep the diffraction from this cause within tolerable limits, the minor axis of an elliptical diagonal should not be greater than approximately one fourth the diameter of the primary mirror.

Extra-focal Diffraction Rings. By moving a high-power eyepiece inside or outside of focus, a bright star image will be seen to expand into a luminous disk which can be resolved into a number of concentric rings, atmosphere permitting. To the initiated eye, these extra-focal rings tell the story of the optics of the telescope. Half a dozen or more may be visible, the outer one being the widest and brightest, and the inner rings diminishing in brightness inwardly (somewhat like a reversal of Fig. 53b). With a perfect mirror, the appearance of the rings will be identical at equal distances either side of focus. (Visual tests will be upset if spherical aberration is present in the eyepiece.) If, outside of focus, the inner rings are brighter than when viewed inside of focus, undercorrection of the mirror is indicated; overcorrection, if a reverse condition is found. In a reflector, a dark area will be observed at the center of the expanded pattern, the innermost rings being obscured by the secondary mirror.

Each ring should appear uniformly bright; if the rings are brighter on one side, the alignment is at fault. At a small distance inside or outside of focus, where only two or three rings are visible, the outer one having considerable thickness, examine closely for any departure from roundness. If the rings are triangular, or have some other odd shape, there is strain or flexure in the mirror. If oval-shaped rings are seen, astigmatism is present, which may be in the eye, eyepiece, or either of the mirrors. Push the eyepiece to the other side of focus, to see if the elongation is then at right angles to the first position — this will relieve the eye of suspicion. If the elongated rings remain unchanged as the eyepiece is rotated in the adapter tube, the eyepiece is eliminated. Rotating the mirror in its cell will then isolate the source of the trouble. A diagonal of poor figure or one pinched in its holder can cause an astigmatic image.

At best focus, one diffraction ring should be visible around a bright star image (Fig. 86). When the seeing is poor, the ring (if at all visible as such) will appear to rotate and perhaps waver to such an extent as to merge with the central disk, at which times the star image may appear as an indefinable splash of light. But with the smaller aperture had by using the diaphragm of Fig. 87, the rings, both in and out of focus, will almost always be seen around any bright star image and, coming from a stopped-down mirror of such relatively high ratio, can be used as a criterion of perfection.[3]

Fig. 86. Appearance of a perfect star image in a reflector. (Linear diameter of the bright ring at the focus of an f/8 telescope is 0.00057".)

The same diaphragm may be effectively used in viewing such trying objects as the full moon or the planet Venus. It is especially helpful for terrestrial observation, permitting the use of a low-power eyepiece with its large field of view, since there is then no objectionable shadow from a secondary mirror, and it also permits of observations on nights when the atmosphere is in such turmoil as to preclude the use of full aperture.

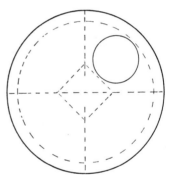

Fig. 87. The offset diaphragm.

Atmosphere, and Thermal Disturbances. The most annoying abomination to plague the amateur astronomer, next to mosquitoes and frostbite, is turbulence in the atmosphere. In general, the atmosphere is built up of layers of air of different temperatures and densities, and therefore of different refractive indices. These layers may stream along like vast invisible rivers in different directions, rippling and undulating, sometimes rising or descending with rushing force. Local topography, nearby mountains, woodland, and the presence of

[3]For further discussions of diffraction rings, see *Amateur Telescope Making Advanced,* and *The Telescope,* by Louis Bell.

large bodies of water exert an influence on these turbulences, which may change from hour to hour. It is small wonder that seeing is frequently poor when one considers that the starlight entering a tele·scope has been buffeted about in its passage through these layers of differing indices. The effects are particularly worsened if the star's altitude is low, the seeing declining rapidly below an altitude of 40°.

Some localities have a large annual percentage of nights when the seeing is good, and high magnifications can be consistently en·joyed. Such places are preferred as sites for observatories. In the arid regions of the West, for example, large telescopes perform bet·ter than along the Atlantic seaboard, where the seeing is not very good. But even the poorest regions in this respect have their share of nights when the atmosphere seems to be perfectly still, and the highest magnifications can be used. In the East, best conditions prevail during the summer months. It is commonplace then, in test·ing the seeing by observing the closer pair (separation 2.5 seconds of arc) of the quadruple star, Epsilon Lyrae, to find the images quite stationary, each disk perfectly defined and surrounded by its diffraction ring. But with the coming of fall, conditions change. In contrast, then, the diffraction disks appear to be doing a dance, and only in momentary glimpses can the diffraction patterns be made out. Hazy nights, while they cut down on the transparency, are usually very good, the tiny droplets of water suspended in the atmos·phere evidently acting as ballast. Especially poor are the transpar·ent "sparkling" nights of winter, when Sirius may sometimes be seen to twinkle in all the colors of the spectrum.

Both refractor and reflector are subject to these observing hin·drances, and there is nothing that can be done about them. But the reflector sometimes has the disadvantage of thermal disturbances inside its open tube. How to control them effectively has long been a problem. Heating units, mechanical ventilation, insulation, and other devices have been tried with more or less success. One form of disturbance is air flowing along the walls of the tube, just as tobacco smoke creeps along a table top. This could be overcome by use of a tube an inch or more greater in diameter than the mirror. But if the tube is taken outside and ventilated about an hour before observations are scheduled to commence, these currents should be swept away. As we shall see, this should be done anyway. It is the practice followed by the author, and I have never encountered any thermal difficulties of this sort inside the tubes of the telescopes shown in the frontispiece and in Fig. 96, both of which are lined with sheet cork.

Properly installed, the sheet-cork lining should be made to hold itself by compression inside the metal tube, without need for retaining rings; the width to which the sheets of cork must be cut should therefore be predetermined by experiment. Consider the case of lining an aluminum tube 56″ long, outside diameter 7″, wall thickness 1/16″, with sheet cork ⅛″ thick. The inner circumference of the tube is 21.6″, and when lined will be reduced to 20.8″, which would suggest that the width of the sheets ought to lie somewhere between these two figures. By experiment with a scrap length of the material, it will be found that the snuggest possible fit can be effected with a width of 21 5/16″. Accordingly, cut to this width two sheets of cork, one 37″ and the other 19″ in length. Locate and cut out holes for the spider arms, eyepiece opening, and so forth, and paint dead black the surfaces to be exposed. Then bend each sheet into a curved "W" shape, and push each separately into its position in the tube, with all edges meeting in butt joints. The bulge should be left opposite the seam, as in Fig. 88, and it is then to be carefully pressed out.

Fig. 88. How sheet-cork lining is fitted inside the tube.

Emphasis has often been placed on the necessity of ventilating the mirror end of the tube at all times, but under certain conditions this does not work out to advantage while actually observing. With the telescope in Fig. 96, heated air rising from the ground has on occasion been found to be passing through the tube, causing a boiling of the image, and this was effectively arrested by fitting a tight cover over that end and keeping it there for the duration of the observations.

The principal reason for giving the telescope ample time for ventilation before observing is to allow the mirror and metal parts to become adjusted to the temperature of the outside air. It has long been known that changing temperature within the substance of the mirror is deleterious to definition. In an effort to observe what takes place while the mirror is adjusting itself to the temperature of the surrounding air, the author experimented with five pyrex mirrors, four of them being excellently figured paraboloids, and one having a spherical figure. These were immersed on different occasions in warm water (96° F.) and cold water (46° F.) for half-hour periods, and subjected to the Foucault test following each immersion, in a room having a temperature of 68° F. These temperature differences, 28° and 22° respectively, are believed to be no greater than may be frequently encountered by the amateur and his telescope. The results of the tests are given in Table III. The third column lists the measured corrections present on the mirrors upon removal from the warm water; the fifth column, the corrections measured upon removal from the cold water. In the fourth and sixth columns are the elapsed times before the mirrors returned to their normal corrections given in the second column.

TABLE III

Size of mirror	Normal corr. (r^2/R)	Cooling	Return to normal	Warming	Return to normal
6″ f/4	0.18″	0.18″	......	0.18″	......
6″ f/8	0.09″	0.14″	25 min.	Spherical	30 min.
8″ f/5	0.20″	0.21″	......	0.20″	......
8″ f/8	0.13″	0.19″	35 min.	0.05″	30 min.
6″ f/9	Spherical	Ellipsoid (0.07″)	30 min.	Oblate (−0.07″)	30 min.

Apparently the figures of mirrors of low focal ratio are little affected by changing temperature, whereas the change in figure of those of high ratio may be pronounced. In testing the f/8's and the f/9, the knife-edge had to be constantly shifted to keep pace with the continually changing figures. And at intervals, far beyond the time required for the figures to return to normalcy, the air in front of each mirror would boil and seethe with convection currents, indicating that for a considerable period satisfactory observing might be difficult.

The findings in the third column of Table III represent the conditions prevalent to a more or less degree in the lowering temperatures of nightfall, when the amateur most frequently undertakes observations, but it should be noted that the difference of some 28 degrees to which the experimental mirrors were subjected was abrupt. In practice, a gradual change of 5° per hour might be encountered, which will hardly affect performance. But sudden changes, as when the telescope is brought out from a warm room, are not conducive to best results.

It might be added, parenthetically, that any zonal errors present on a mirror (above f/5) will stand out in bold relief if subjected to the hot-water treatment and then given the knife-edge test.

Because of the larger cylinder of air through which the light from a star must pass to reach the mirror, a large telescope has little chance of performing up to the theoretical limit of its resolving power. It is frequently found necessary, in order to continue observations, to diaphragm large telescopes down to a smaller aperture and thus to diminish the effects of atmospheric turbulence.

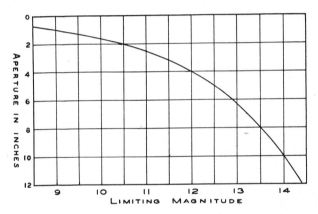

Fig. 89. The minimum visible magnitude per telescope aperture.

The Limiting Visual Magnitude of a Telescope. The early Greek astronomers, Hipparchus and Ptolemy, classified the stars in order of brightness. The brightest stars were designated as 1st magnitude, and those just bordering on the limit of visibility as 6th magnitude. John Herschel found that a 1st-magnitude star was on the average about 100 times brighter than one of the 6th magnitude, and that there was a geometrical progression in the scale, stars of any given magnitude being about 2½ times as bright as those of one magnitude lower. This ratio has now been fixed at the number whose logarithm is 0.4, or very nearly 2.512. A standard 1st-magnitude star ($1^m.0$) is 2.512 times brighter than one of $2^m.0$, 2.512^2 (or 6.3) times brighter than one of $3^m.0$, 2.512^3 (15.8) times brighter than one of $4^m.0$, and so on. On this scale, a standard 1st-magnitude star is about 25,000 times as bright as one of the 12th magnitude.

The limiting visual magnitude of the naked eye, with a pupillary aperture of about 7 mm., is somewhat better than $6^m.0$ (some eyes can detect stars of $7^m.0$). Since the light grasp of a telescope is proportional to the square of its aperture, a 1-inch telescope should therefore reveal stars of about the 9th magnitude. On this basis, which has been generally adopted, Fig. 89 has been prepared, showing graphically the approximate limiting magnitude for telescopes of different apertures.

Aberrations of the Paraboloid. Definition on the axis of your paraboloidal mirror should be limited only by diffraction. That is the property of the paraboloid, and at any focal ratio perfect images will be found on axis. Used photographically, for the recording of

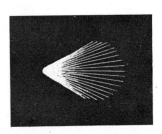

Fig. 90. Comatic star image.

nebulae and clusters, seldom exceeding in diameter about 20 minutes of arc, your mirror should give practically perfect results. This is because the diameter of the image of a star on the plate within that limited field will easily exceed the size of the off-axis aberration image. But the farther off axis the image is placed, the more it is affected by the aberrations discussed in this section. The considerations presented here should be carefully weighed in planning a second telescope (see the next chapter) or in adapting the first one for photography. The principal advantage of a wide-aperture telescope lies in its light-gathering power, and if it is of low focal ratio its speed makes it most useful for photographing faint nebulae and clusters.

The most serious off-axis aberration is coma, which gives star images an "opened-parachute" shape, pointing toward the axis (Fig. 90). The effect of this aberration increases with the distance from the axis, and as the inverse square of the focal ratio. (The amount of image spread can therefore be lessened by diaphragming the mirror, and so decreasing its angular aperture.) The length and width of the comatic image are in the ratio of 3 to 2, with the light concentrated near the point.

Of secondary importance is the astigmatism of off-axis images (not to be confused with astigmatism from an unsymmetrical mirror,

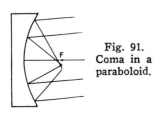

**Fig. 91.
Coma in a
paraboloid.**

which affects axial images as well), the aberration increasing with the square of the image distance from the axis. Coma is more pronounced at moderate and short focal ratios, but at f/15 or longer astigmatism predominates. Whereas in coma the images formed by the different zones of the mirror are not concentric (as shown in Fig. 91), astigmatism arises from the failure of rays from different parts of the same zone to intersect in the same plane. In the absence of coma, astigmatism produces an elongated or linelike star image on the focal plane.

The aberrations of a paraboloidal mirror (unlike those of a lens) are completely fixed by the choice of diameter and focal length. The combined effect of all aberrations can be shown roughly by spot diagrams like those in Fig. 92. A computer was used to trace dozens of incoming parallel rays striking evenly spaced points on the mirror, in order to find out where, after reflection, they would intersect the focal plane. Each cluster of spots corresponds to a star's image, as formed at various distances from the optical axis. Where the spots are closely packed, the image is most intense, while parts where they are widely spaced represent low, possibly invisible, light levels.

Each clump of spots is reproduced at the same linear scale, magnified to nearly 300 times actual size (the bar at lower left shows 0.001 inch). The three rows represent star images formed by 6-inch f/12 (top), f/8 (middle), and f/4 mirrors. In each case, the on-axis spot at left is surrounded by a circle the size of the 1.8-second-diameter diffraction disk for a 6-inch mirror (see page 172). To the

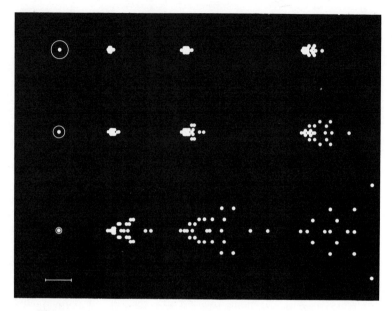

Fig. 92.   Spot diagrams for 6-inch f/12, f/8, and f/4 paraboloidal mirrors, as calculated by Roger W. Sinnott.

right of this circle, in order, are spot images for the edge of
¼-, ½-, and 1-inch-diameter fields of view (the latter being the
limit for ordinary eyepieces). Because of the different image scales,
a 1-inch field corresponds to 48 minutes of arc with the f/12 mirror,
1° 12′ with the f/8, and 2° 24′ with the f/4.

It is seen that the field of good definition of the paraboloid
is rather small, and extremely limited in an f/4. The image will seem
very nearly perfect, however, if most of the individual spots fall
within the area of the ever-present diffraction disk. For a 6-inch
f/8 mirror, this occurs only within a ½-inch field of view (barely
taking in the moon). Obviously, alignment of the mirror must be
good enough to place the mirror's optical axis squarely in the middle
of this pea-sized central region of the eyepiece drawtube, if the
best performance is to be realized.

Chapter XIV

A SECOND TELESCOPE

OFTEN, BEFORE HIS FIRST TELESCOPE IS FINISHED, the enthusiastic amateur is planning on a second one. This is the usual manifestation of a healthy interest in an exacting and fascinating hobby, but it should be subjected to a modicum of restraint. The tyro should be wary of any impulse to make a telescope of great aperture and focal length, and should carefully consider the practicability of such a project, and the probability of its successful conclusion. He must also consider the ways, if any, in which a second instrument can supplement the observing program planned for the first telescope.

Pyrex mirror blanks are cast in stock sizes of $4\frac{1}{4}''$, $6''$, $8''$, $10''$, $12\frac{1}{2}''$, and $16''$, providing a range of apertures calculated to satisfy the most intrepid individual. But large mirrors (10-inch and more) must be figured to a most rigorous tolerance; also, the larger the mirror, the greater the volume of atmosphere with which it must cope. The mounting parts may assume formidable proportions, and require an extensive amount of machining. Comparative costs of telescopes of different sizes have been estimated to vary with the cube of the aperture, although many amateurs have ingeniously solved the mechanical problems and materially reduced the expense by the adaptation of secondhand automobile parts for mountings. And in order to observe comfortably while standing on the ground (the height of one's neighbors should also be considered), the height of the eyepiece position should not in general exceed about $60''$ when the telescope is pointing to the zenith. This places a practical limitation on the focal length of the mirror, unless at times one is prepared to put up with the inconveniences of a portable platform or a stepladder.

If the demands on one's optical skill are not to be too severe, the telescope is pretty well restricted, in the $4\frac{1}{4}$-inch size, to ratios above f/6, and in the 6- and 8-inch to ratios of f/6 to f/8. These

185

proportions permit slight departures of the figure from a paraboloid, without material impairment of the definition. The beginner can also choose from these sizes, if for some reason he does not elect to make the 6-inch f/8.

In lesser ratios in the above diameters, and in all ratios in the large sizes, any measurable departure of the figure from a paraboloid cannot be considered, as the allowable tolerance in knife-edge movement will then be exceeded by the error of the testing observation.

Long-focus Telescopes. If you wonder why the 6-inch f/9 and f/10 are omitted from the list of "easy" mirrors, it should be stated that unless one has already made a pretty fair mirror, and has become very familiar with the parabolizing stroke and the shadow appearances, he will in all likelihood not succeed with an f/9 or f/10. It has often been said that an f/10 can be finished with a spherical figure, but the spherical aberration from such a mirror seriously impairs fine contrasts. Only a casual worker will be content with inferior performance.

Figuring these nearly spherical mirrors is quite difficult. Under test, the paraboloidal shadows appear so faint and delicate that to the uninitiated eye they practically defy analysis. Figured to a stage where the boundaries between light and shadow become clearly definable, it is a certainty that there will be strong overcorrection. In parabolizing a shallow mirror, best done on a hard lap, there is no time for experimentation with the stroke, so quickly is the figure altered. The proper stroke must be known and used at the start, and this knowledge can be gained only through previous practice. But the patient and able worker is amply rewarded in the exquisite definition to be had from these mirrors of longer focus.

By the simple expedient of hurdling the f/10, however, and making a 6-inch mirror of ratio f/12 or more, the amateur avoids the necessity of parabolizing, and can finish the surface with a spherical figure. He will achieve the most perfect long-focus reflector performance that it is possible to obtain, and have a telescope particularly suited for lunar and planetary observations.

Since the f/10 and higher-ratio instruments are primarily intended to be used with high magnification, their diagonals need be only sufficiently large to provide full illumination over the field of a $\frac{1}{2}''$ eyepiece. This calls for a width of about $1''$ in the case of the 6-inch f/10 mirror, and $\frac{3}{4}''$ for the f/12. The diffraction effects from such proportionately small diagonals should be quite invisible. It may prove difficult, however, for the amateur to construct a supporting mechanism that will not extend beyond the boundaries of a diagonal of such small size. It might be more practical, rather than using the hinge and spring of Fig. 59, to make provision for the angular adjustment by slotting the holes in the telescope tube where the spider is anchored. Then, by attaching the diagonal rigidly to its support, at as near a 45° angle as possible, any discrepancy can be corrected by tilting the entire spider in the slots. In this method, the position of the spider should also be changed from the one shown in Fig. 63, the vanes being placed in vertical and horizontal planes (with respect to that diagram). Only those holes lying in the vertical plane, or in the plane of the eyepiece, need be slotted.

The longer tube makes the mounting more sensitive to vibrations from air movements; therefore, if a pipe mounting as described in Chapter X is going to be constructed, the type shown in Fig. 73 should be the choice.

Short-focus Telescopes. An ideal companion telescope for a 6-inch f/12 is a 6-inch f/4 which, equipped with a low-power eyepiece, will embrace a field of view slightly less than $2\frac{1}{2}°$ in diameter. This rich-field telescope (*RFT*) furnishes beautiful views of rich star fields, clusters, and nebulae. With it, the Andromeda galaxy can be seen extending over the entire width of the field; it is also possible to distinguish the North America nebula in Cygnus. Thus armed with two instruments of equal light grasp, and a possible range in powers from 21x to 240x or more, the amateur is well equipped for a most varied program of astronomical study.

In the discussion of exit pupil in Chapter IX, it was shown that the least magnification which should be used on a telescope is one that yields an exit pupil of about 7 mm., found to accompany a magnification of about $3\frac{1}{2}$ per inch of aperture. This corresponds to a power of 21 for a 6-inch telescope of $24''$ focal length, and in turn requires a focal length of $1\ 1/7''$ in the eyepiece. That is a rather odd focal length, not ordinarily obtainable; therefore, the focal length of the mirror should be so adjusted that its combination with a good low-power eyepiece of whatever focal length can be had, will furnish a 7-mm. exit pupil. If you propose to use a 1″ eyepiece, then the focal length of the mirror should be 21″, but this combination will result in less than a 2° field. Moreover, because of increasing aberrations and the difficulty of figuring, it is not advisable to go below f/4. A superior combination, embracing a field of about 2°, is an eyepiece of 1½″ focal length and a mirror of ratio about f/5. A Kellner eyepiece of about 1¼″ focal length, with a 50° apparent field, should not be too difficult to obtain, and used with a mirror of about 26″ focal length, it will probably give most satisfactory results.

We have intimated that a low-ratio mirror must be very precisely figured, and this is true if it is to perform within the Rayleigh limit or even within twice that limit. But in the size, and for the wide-angle observing of which we are speaking, there is no necessity for correcting the mirror to those exacting tolerances. Primarily, a wide field of view, some 2° or more in diameter, is what is sought, and even a perfect paraboloid of the proportions given cannot combine all of the rays to meet within a half wave for a star more than a few seconds of arc from the mirror's axis.[1] And with the low powers that will be used, the observer can hardly become aware of any deficiency that may exist near the center of the field. But no leniency can be extended to the diagonal, which must be flat in any case. A poor diagonal will do more injury to the images than will a poor mirror.

For all deep mirrors (small focal ratio), the pitch lap should be somewhat on the soft side, as considerable flow is demanded in following the deepening curve of the mirror as its figure approaches the paraboloid. And here one must have the patience to carry on the figuring strokes for comparatively long intervals, in order to give the lap a chance to function. A deep mirror does not necessarily have to be brought to a spherical figure before commencing the parabolizing. This practice is pursued for the shallower mirrors because the sphere provides a fix, to borrow a navigator's term, close to the final destination, and in the short remaining run one is not expected to stray far from the course. But with an f/4 or f/5,

---

[1]See the discussion on *Aberrations of the Paraboloid* in Chapter XIII.

so much work remains to be done that it is probably a waste of time to struggle for the sphere, and the parabolizing strokes, described in Chapter VI, should be adopted as soon as the polishing shows signs of completion.

On a deep mirror, there is ample time to discover any existing zonal discrepancies by means of the knife-edge test, and to correct them by experiment with the strokes. A reasonably good figure should be present on the mirror when it is judged to be completed. Perhaps, after using the telescope for a time, and enjoying the benefits of his labors, the amateur will decide to refigure the mirror as close to perfection as possible, and then try to determine if any improvement in its performance is discernible.

Fig. 93. Featherweight Newtonian, 3-inch f/4, made by Jack Smollen, Hurley, N. Y. The design of the bearings is similar to that of Fig. 71.

A low-ratio mirror is very sensitive to misalignment; the slightest error will produce an annoying flare on a star image. Therefore, the utmost care should be taken in the construction of the cell and other parts. For improved accuracy, try 1/16″ instead of 1/8″ holes in the cardboard disks used in aligning. For a 6-inch f/4, the offsetting of the center of plate E, Fig. 59, ought not to be ignored, as in this case it amounts to about six per cent of the minor axis of the diagonal. With this offset, you may find it somewhat disconcerting when reflections cannot be brought concentric as you peer through the adapter tube while attempting to adjust the angle of deflection. This condition can be partly overcome by making the diameter of the diagonal larger by the amount necessary to effect centralization in the tube, but as it already obstructs a considerable percentage of the light, any further increase should be avoided. Therefore, a disk of blackened cardboard of the larger diameter required can be temporarily placed on the end of the rod *H*, Fig. 59, and discarded after the adjustments have been completed. On out-of-focus star images, of course, the shadow of the diagonal will not be quite concentric with the diffraction rings, but as long as the visible rings are round, and each one evenly bright, the adjustments must be in order.

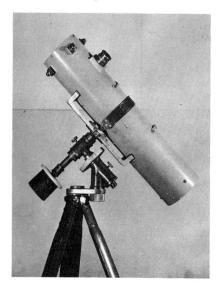

Fig. 94. A 4¼-inch f/4.5 portable Newtonian, made by the author. Note the offset of the mounting, making possible omission of a counterweight on the polar axis.

The f/4 can, if necessary, be more lightly mounted than the f/8 or f/10, due to its much shorter tube and the lower magnifications that will be used with it. Fittings one size smaller than those listed under "A," Table I, will be satisfactory; for instance, the axes may be of ¾" pipe, and the tees 1" pipe size. Also, because of the shorter tube, the mounting can be attached directly to the tripod head, dispensing with the pillar. (See Fig. 93.)

A 4¼-inch Telescope of Long Focus. Almost everything that has been said of the 6-inch can be applied to a 4¼-inch, a size of reflector which seems to have been considerably underrated. It is the regrettable truth that vast numbers of large and excellent telescopes are collecting cobwebs in attics, garages, and cellars, simply because they are too massive to be readily moved outdoors and in. No doubt their owners were once enthusiastic amateur astronomers, but their ardor has been dampened, and they have been denied many anticipated pleasant evenings of observation, because a too-ambitious program was undertaken as an initial venture. The 4¼-inch reflector is a powerful instrument in its own right, far more telescope than Galileo ever had. Including a tripod, its total weight is trifling, between 20 and 30 pounds. At f/10 and over, the mirror can be finished spherical. If you are after magnification, this size can be made f/15, at which ratio it will compare in performance to a refractor of similar dimensions. The parts for the mounting of a 4¼-inch, of ratio above f/6, should be of the size listed under "A," Table I.

Larger Instruments. A telescope larger than the 6-inch can hardly be made portable, as that term is generally understood. Of course, the mounting can be placed on a tripod, to permit moving it about, but the handling may not be easy. A possible exception is an 8-inch f/4 or f/4.5, which can be carried on the mounting shown in Fig. 73.

The selection of fittings for larger and permanently mounted telescopes is left to the discretion of the amateur, but remember that

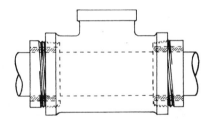

Fig. 95. Short nipples made up tight in a tee fitting increase the bearing length. This method was used in the construction of the mounting shown in Fig. 96.

the larger they are, the better. In Fig. 95 is shown how more efficient use can be made of the larger size fittings needed for an 8-inch or 10-inch telescope. Two short nipples (2½" pipe size), threaded at one end only, are made up tight in a 2½" tee. The axis is 1½" (pipe size), or of solid stock, 1⅞" in diameter. The babbitt bearing is housed in the nipples. In pouring, the assemblage is stood at a steep angle, and each half poured separately through the branch opening. Or, a bearing may be poured in each nipple before making up, around a smaller diameter shafting, and, after making up, the entire bearing can be bored out in the lathe to the required diameter. The long span and large section area combine to make this an extremely rigid assembly. Bearings of similar design are used in the 8-inch telescope mounting illustrated in Fig. 96.

For roughing out the curve on moderate-sized mirrors, No. 80 carbo may be considered about as fast as the coarser No. 60, and it does not result in as deeply pitted a surface. But for 8-inch mirrors of f/6 and lower ratio, and on all large mirrors, it will be more economical to do the excavating with No. 60. Grinding may then proceed with No. 80, or a series consisting of Nos. 100, 150, then 220, and so on may be used. As the amount of pressure per unit area of a large mirror can hardly equal that used on a 6-inch, it will take a longer time to grind down each charge of abrasive, and at least the same number of charges of each grade must be used.

Fig. 96. An 8-inch f/8 Newtonian, with babbitted bearings. Axes are of 1⅞″ solid shaftings; the 10″ tube is of 22-gauge sheet metal; saddle and rotation rings are of aluminum. This instrument, made by the author, is rigid and well balanced.

Testing Notes. On large mirrors, it is usual to supplement the methods of testing already described with measurements made in several zones. For this purpose, diaphragms with ½″ zonal openings (Fig. 97) are used, the mean radius of the openings being taken as the radius of the zone being tested. The central aperture in the mask in Fig. 38a is one quarter of the mirror's diameter, which is satisfactory for a 6-inch f/8 mirror or higher ratio, but on larger and deeper mirrors this opening should not exceed 20 per cent of the diameter. Zonal testing is somewhat tedious, but nevertheless necessary, as there is too great an area on large mirrors to depend on the eye detecting minute vagaries in shadow appearances without some means of isolating them.

As a useful supplementary test, two or three disks of cardboard may be made, each with a different-sized central opening, the smallest being, perhaps, equal to half the diameter of the mirror. As each such stop is in turn placed before the mirror, a different size of "mirror" of different focal ratio is exposed, and the paraboloidal shadows of each can be examined. If desired, use separate masks, similar to Fig. 38a, for each aperture. Zonal discrepancies will show up under this test about as readily as by the method mentioned above. At corresponding settings of the knife-edge, the figures should have similar appearances, except for a difference in the depth of the shadows.

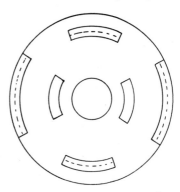

Fig. 97. Type of zonal mask used in the testing of large mirrors.

The shadows on an f/4 or f/5 paraboloidal mirror are quite intense, looking somewhat like those of Fig. 33 when the knife-edge is placed at the 50-per-cent setting. Seen on a 6-inch f/8, shadows of this intensity could only accompany a deep hyperboloid. But on an f/8 mirror of larger diameter, where the correction is greater than on the 6-inch, the contrast between highlight and shadow also increases. For a 12-inch f/8, with double the amount of correction of the 6-inch, Fig. 33 might well be an accurate representation of its appearance when parabolic.

Tolerable Knife-edge Error. It has been found that reasonably perfect performance is assured if the mirror's surface does not vary by more than a quarter of a wave length from a paraboloid. The allowable amount of undercorrection or overcorrection, or the departure in inches from the value of r^2/R, for mirrors so figured, is given in Table V. Note that the allowable error varies as the square of the focal ratio. It is independent of the diameter, and therefore applies to mirrors of any size.

TABLE V. TOLERABLE ABERRATION ERROR IN PARABOLOIDAL MIRRORS WITH STATIONARY PINHOLE AT CENTER OF CURVATURE

Focal ratio	Error in inches	Focal ratio	Error in inches
f/3	0.006	f/7	0.035
f/3.5	0.008	f/8	0.045
f/4	0.011	f/9	0.057
f/4.5	0.014	f/10	0.070
f/5	0.018	f/11	0.087
f/6	0.025	f/12	0.101

For Newtonian telescopes, the above theoretical tolerances may be taken to the nearest hundredth of an inch, but because of probable testing errors, which may amount to as much as 0.02″, it is advisable to reduce the allowable knife-edge error where possible by up to that amount. It is evident therefore that mirrors of f/6 and lesser ratio must be very precisely figured. (See, however, considerations mentioned on page 188.) High-ratio mirrors, for which the tolerable aberration error exceeds the value of r^2/R, may be left with a spherical figure. For example, the aberration at focus of a spherical mirror may be very closely approximated from the value of $r^2/4R$, which for a 6-inch f/11 will amount to 0.017″. But if, in figuring, full advantage were taken of the tolerance given in Table V, the mirror's aberration at focus would then be 0.022″ (0.087/4). The spherical 6-inch f/11 mirror is thus seen to lie well within a quarter wave of a paraboloid.

Appendix A

SECONDARY REFLECTIONS

Disks of Unequal Sizes. It sometimes happens that the tool and mirror blanks may not be of exactly equal diameters. A difference of about ⅛″ does not matter, as the working surface of either disk can be beveled by whatever amount is needed to make them both equal.

Broken Disks. Occasionally, before completion of the mirror, either it or the tool may be broken. If this happens to the tool in the polishing stage, a flat iron disk or another glass disk can be obtained, and a new lap made on one of its flat surfaces. The greater thickness of pitch at the center will introduce no difficulties, and polishing and figuring can be successfully concluded.

If the tool is broken before grinding is completed, the mirror must be ground with No. 80 on top of the new tool, using one-half diameter strokes, until the grinding reaches the center of the tool, when the radius of the mirror should be measured, and the procedures outlined in Chapter III then applied.

If it is the mirror that is broken, rough grinding of a new blank on the already curved tool may begin with half-diameter strokes, in which the center of the mirror traverses a zigzag course like that shown in Fig. 30. If the resultant curve becomes too shallow, increase the amount of work over the edges of the tool; if too deep, work more over its center. When the curve has been extended to about 1/16″ from the edge of the mirror, resort to one-third strokes to bring the surfaces spherical.

Concrete Suggestions. The concrete for the base of the grinding stand (Fig. 17) should be a rather stiff mix, of parts 1, 2, and 4, of cement, sand, and gravel. Keep it moist for several days while curing. The base for the concrete pier for a permanent mount can

195

be of a leaner mix, 1, 2, and 6. Brick and stone can be used as a filler. The pier, of course, should be of the stronger 1, 2, and 4 mix.

The bolts holding the mounting to the pier can be quite accurately bedded in the concrete in this way: Transfer the holes from the heavy flange (*A*, Fig. 72) to each of two plates of sheet metal, one of which has been cut away to a skeleton shape. First pass the four bolts through the corner holes of this skeleton plate and screw nuts onto the threaded ends. Fit the other plate down against the nuts, apply four more nuts, and lock the second plate tight. Then set the assembly, skeleton plate down, in the wooden form and orient it carefully as explained in Chapter XII. After the concrete has hardened, the upper plate and the nuts can be removed, and the mounting can be set down.

Blind Flanges. As a rule, the holes found in heavy flanges are much larger than are needed, so a blank or "blind" flange should be asked for. If the dealer does not have it in stock, let him order it. Holes of the correct size can then be drilled, but they must be slightly larger than the bolts (which might be $5\!/\!8''$ in diameter), to permit tilting for latitude correction.

Springs. For the mirror cell and other parts, if compression springs of just the right length cannot be found, longer ones can be cut off by holding a coil against the corner of a grinding wheel.

Working Time. The question is often asked, "How long does it take to make a mirror?" This is largely a matter of experience. A woman student finished her first mirror in the exceptionally fine time of about 25 hours. Confronted with the necessity of replacing a broken mirror, the author completed a 6-inch f/8 in two evenings, a total of about eight hours' working time. On the other hand, mirrors have been known to take two years or more. Figuring can well consume the bulk of the time put in. The curve should be achieved in two or three hours, and allowing a liberal hour for each of the stages of fine grinding makes a total of eight hours for grinding. Lap making, one hour, and six hours for polishing, brings the total time up to 15 hours. Figuring, *x* hours. The average amateur, working alone, from the directions given in this book, should complete a good mirror in from 30 to 40 hours of working time.

Making the diagonal is a much simpler proposition. Here one is concerned only with the small piece of glass at the center of a large disk. Thirteen diagonals, cemented to an 8″ disk, and ground and polished in conjunction with two other disks in the manner described for making optical flats (Appendix B), were completed by the author in a total of about 10 hours, the surfaces of the diagonals being flat to a quarter wave or better.

Observing the Sun. Never attempt to view the sun directly through the telescope without first doing something about the terrific concentration of heat and light, else the penalty may be a hole burned in the retina, perhaps permanent blindness. This warning can be modified to the extent that the setting sun, when it appears dull and coppery, hanging just above the horizon, may be looked at with the aid of a piece of smoked glass or other filter held between the eyepiece and the eye. When it is higher in the sky, the sun may be observed only when complete protection has been afforded the eye.

There are a variety of ways of reducing the sun's light by the required millionfold before it may be safely viewed. Filters, if used, should ideally keep most of the light and heat from ever entering the telescope. Full-aperture filters are sold as telescope accessories, but the cost is high because the glass must be finished to stringent optical standards. These filters utilize vacuum-deposited metallic coatings to cut transmission of all wavelengths, including the infrared. Filters of lower optical quality can be tolerated if they are used near the focal plane, but the sun's heat is greatly concentrated here, and a dense solar filter is very liable to crack under these conditions. The hazard is lessened by stopping the mirror down and using several filters; then, if one cracks, the others provide momentary protection. Use only welder's glass filters for this purpose, as they are specially formulated to exclude harmful infrared and ultraviolet rays. The combination should equal in density a shade No. 13. Disks can be biscuit-cut on a drill press to fit the eyepiece drawtube.

A stop such as that shown in Fig. 87 can be used, although the resolving power is limited to that of a telescope having the same aperture as the opening. Another way of diaphragming the mirror is to secure a 5″ cardboard central stop to the rod supporting the diagonal, thus leaving exposed a ½″-wide annular zone.

An Exclusively Solar Telescope. But if the telescope is to be used only for solar observations, the full resolving power of the mirror can be retained by coating the primary and secondary mirrors, not with a reflecting substance, but with the anti-reflecting magnesium fluoride. The net result is that only a fraction of one per cent of the light striking the main mirror reaches the eyepiece, but a few filters may still be required. The back surface of the diagonal should be rough ground and painted black to cancel out any reflection from that source. Better still, instead of the ordinary diagonal, is the Herschel wedge. As its name implies, this is a wedge-shaped glass prism devised by Sir John Herschel; its back surface is so inclined that any reflection from that source is pitched out at a radically different angle to that from the first surface, and so it does not interfere with the vision. If a right-angle prism is used for the secondary reflection, its hypotenuse or reflecting surface should face outward, so as to reflect externally.

A Too-bright Moon. Lunar observation, ordinarily rather trying on the eyes, and at times very much so, can be rendered more comfortable by the introduction of some of the filters or stops described above.

Sunspot Projection. Sunspot observation can be enjoyed by several persons at the same time simply by projecting the sun's image onto a white cardboard screen. The eyepiece (a low power is best) is used as a projecting lens by drawing it slightly out of focus. By changing the eyepiece position and adjusting the distance of the screen to suit, the size of the projected image can be varied. This method is awkward unless a support for the screen is made. A more satisfactory device is a box, light-tight and light in weight, that can be strapped to the telescope straddling the eyepiece, the bottom or outer surface of the box being formed of a sheet of ground glass onto which the sun's image is projected. Means of access should be provided for adjusting the eyepiece.

A Driving Clock. The advantages of an equatorial mount can be fully realized by the addition of a driving mechanism to the polar axle. Old alarm clocks, falling weights, and even water dripping from a tube have all been harnessed by ingenious amateurs, but the best source of motive power is undoubtedly a synchronous motor, whose turning rate is accurately governed by 60-hertz house current. Through a gear train it turns a worm and worm wheel mounted on the southern end of the polar shaft (see Fig. 98).

The gear train will be simpler if the motor itself runs slowly. Synchronous motors are available for 1 r.p.m., 1/15 r.p.m., 1/60 r.p.m., and a variety of other rates. Often, such a motor's shaft can be attached directly to the worm.

For visual observing, it is sufficient to drive the polar axle at the solar rate, one revolution in 24 hours. Assuming, for example, a 1-r.p.m. motor, the reduction required is 1/1,440. Let us say that we have a worm wheel with 100 teeth. A worm is equivalent to a gear of one tooth, so the worm and worm wheel act to slow the turning rate by 1/100. The remaining factor of 1/14.4 can be accomplished by choosing a spur gear of 10 teeth for the motor shaft (C in Fig. 98) and one of 144 teeth for spur B. Alternatively, if a motor of 1/15-r.p.m. is used, it may turn the worm directly, in which case a wheel of 96 teeth should be obtained.

But if the instrument is to be used for astrophotography, it is preferable to use the true sidereal turning rate. In his

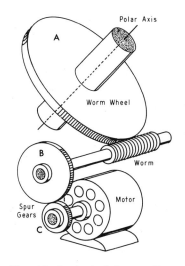

Fig. 98. A clock drive configuration by E. A. Fagen.

study of the problem near the beginning of this century, E. S. King of Harvard Observatory concluded that, because of refraction effects, the best compromise in driving rate is one that loses 24^s in the $23^h\ 56^m\ 04^s$ sidereal day. In other words, the axle should turn once in 86,188 seconds, or about 1436.5 minutes. This need not be achieved *exactly*, because small corrections, or "guiding," will still be necessary in most parts of the sky. Approximations to this rate, using a 1-r.p.m. motor, can be obtained with A = 252 teeth, B = 57, and C = 10; also, with A = 338, B = 170, and C = 40. In any case, the choice of one of these, or perhaps of trains with more gears, will be dictated by the availability of gears with uncommon numbers of teeth. Naturally, meshing gears must be of the same pitch.

A convenient way to mount the worm wheel is illustrated in Fig. 99, where a friction clutch permits the telescope to be swung

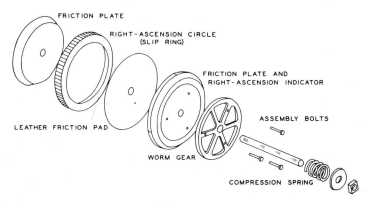

FRICTION PLATE

RIGHT-ASCENSION CIRCLE
(SLIP RING)

FRICTION PLATE AND
RIGHT-ASCENSION INDICATOR

ASSEMBLY BOLTS

LEATHER FRICTION PAD

WORM GEAR

COMPRESSION SPRING

Fig. 99. An exploded view of the wheel assembly,
drawn by Owen Gingerich.

to a new object without disengaging the drive. When the telescope is not actually being slewed, there is enough friction between the leather pad and the metal friction plates to drive the telescope, the compression spring being adjusted for the best effect. The right ascension circle must be set (by means of a bright star) at the start of an observing session; thereafter, as long as the drive is running, an indicator on the lower friction plate (attached to the worm wheel) will point to the right ascension of any object under view, without the need to calculate sidereal time.

Fluoride Coatings. When a hard permanent fluoride coating is applied to optical surfaces, the temperature of the glass must first be brought up to about 400° Fahrenheit. This cannot be very quickly done, and is quite a slow process in the case of thick mirror disks. Some technicians may be unwilling to wait the required time, in which event the fluoride coating will be too soft, affording no protection to the aluminized surface, and coming off with the least bit of rubbing. Unless assurance can be had in advance that a durable coating will be furnished, the money spent may be wasted. If the job is properly done, the coating should be able to withstand vigorous rubbing with soft cotton or tissue. There should be no excuse for failure with the secondary mirror, however, which is as easy to treat as the average lens or prism, and as the secondary is at the dewy and dusty end of the telescope, it should be good economy to fluoride-coat its aluminized surface.

Tying the Tripod Legs. To prevent the tripod from collapsing, the legs must be securely linked together. One method is to run separate lengths of chain or flexible cable from leg to leg, or to each leg from a common center. An alternate method is to use rigid members of wood or metal, either joining leg to leg or running to a common central junction. The flexible ties are easiest to install, but rigid ones will lend added stiffness to the tripod.

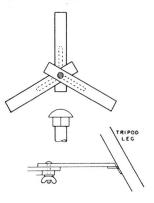

Fig. 100. The tripod links.

An excellent method of securing the legs is shown in Fig. 100. The hardwood ties, of suitable length, and about 3/8″ thick by about 1¼″ wide, are joined to the legs by ordinary hinges. A 3/8″ carriage bolt (see inset) and wing nut lock the three ties in a common union at the center. The spread of the legs can be varied if slots, indicated by the dotted lines, are sawed or milled in the ties. Since the latter are each in different planes of elevation, they must be hinged to the legs at appropriately spaced intervals, and the leg attachments must be designed with this end in view.

Appendix B

MAKING AN OPTICAL FLAT

THE METHOD DESCRIBED IN CHAPTER VII for grinding and polishing the diagonal is not the one usually pursued in the making of an optical flat, although it is capable of a high order of accuracy. Of course, any means of grinding that will result in an apparently perfectly flat surface, when tested against a high-quality straightedge, should be sufficiently reliable to warrant polishing, in the expectation of producing a perfect plane. A master flat of known precision is essential, however, if the operation on a single surface is to be successfully concluded.

When a master flat is lacking, it is customary to proceed by the three-disk method for the operations of grinding, polishing, and testing. Three glass disks of equal diameter, and each not less than one eighth this diameter in thickness, are marked for identification, and are then ground together in a logical sequence, so that each of them will be accorded an equal amount of work against each of the others in every possible combination, the endeavor being to produce an infinite radius on each of the surfaces. The sequence of the grinding is as follows: *a* on *b*, *b* on *a*, *a* on *c*, *c* on *a*, *b* on *c*, *c* on *b*, *a* on *c*, *c* on *a*, *a* on *b*, *b* on *a*, *c* on *b*, and *b* on *c*. The identifying letters can be painted in one or more places on the sides of the disks.

One charge of abrasive is thoroughly ground down in each position, and the whole sequence should be gone through with each grade; thus each surface has the benefit of eight charges. Quarter-diameter strokes should be used throughout. As when fine grinding the mirror, occasionally separate the disks and redistribute the abrasive with the fingers. The heavy pressure that is normally maintained in grinding should be relaxed for the last half of the sequence when using No. 600, and little if any pressure should be used on the final grade. Ample grinding with each grade of abrasive is afforded each surface, so do not spend time looking for pits or testing with a straightedge. Properly worked, each of the surfaces should emerge flat to a wave length of light or less, as disclosed by the interference test when the disks have been sufficiently polished to yield a reflection. (See Chapter VII for a discussion of the light source and interference test.)

The surfaces of plate-glass disks are manufactured reasonably plane, so if they are to be used for the flats, grinding can begin with No. 600. If pyrex, by far the best material (except for fused quartz), is used, the grinding must commence with No. 80 carbo. The back surfaces of pyrex mirror blanks are surrounded with a wide roughened border that makes difficult the viewing of the inter-ference bands out near the edge, so it is customary first to grind and polish those surfaces to a reasonable plane, the rough grinding being best done on a large slab of scrap plate glass.

It is desirable to use two laps in polishing, one being pressed while the other is working, with the best flat used for pressing. Plate-glass tools of suitable thickness can be used for the lap foundation, and the pitch must be on the hard side; in fact, it is difficult to have the lap too hard for flat making. The molded lap with evenly spaced facets (see Chapters IV and VII) is admi-rably suited for this work, as it is markedly superior to the channeled lap in freedom from zones. For the polishing, one-quarter to one-third diameter strokes should be used. Polishing periods may be 15 or 20 minutes long, followed by at least half an hour of cold-pressing. About 30 pounds of weight should be used in pressing, equally distributed at three points on the top of the flat, and since neither the bottom of the tool nor the top of the bench will be per-fectly flat, a thin cushioning material, such as sheet rubber 1/16″ thick, should be placed between them. Cerium oxide or Barnesite, used as the polishing agent, will materially speed up the work.

Convexity is corrected in the normal way, with the flat on top, and by extending the stroke length to as much as half-diameter. A small amount of concavity, of the order of a wave length or so, can be corrected in the same way, but using short strokes; for greater error it may be necessary to invert and polish with the flat face up. There will be little danger of turning the edge by this procedure, provided the lap is sufficiently hard and the strokes are kept short. Frequently, a surface consisting of a compound curve will crop up;

in other words, a real cross section like that shown in Fig. 32c will be present, the center of the flat being convex, and the edge zones concave. Except in a very mild form, such a condition is rare with a hard lap. It is corrected by polishing with the lap on top, using one-third strokes, the effect being to throw the whole surface convex, after which the flat is placed on top, and the convexity reduced with as long a stroke as can be handled without introducing concavity at the center. After testing, and before resuming polishing, you must first establish contact by means of hot-pressing, allowing the flat to stand on the lap for at least one hour, under 30-pound pressure.

On account of temperature effects, far more pronounced than in the case of the concave mirror, figuring will have to proceed slowly. For a quarter-wave tolerance, the relatively liberal conditions prescribed for mirror making are satisfactory, but as the figure approaches to within 1/10 of a wave of flatness, as much as three hours may have to elapse, following an interval of polishing, before an accurate test can be made. During this period, the room must be free from drafts, and the temperature should not vary by more than one degree. Do not handle the flats any more than is necessary to place them in contact for testing and to space the bands properly.

For best accuracy, space the bands about 1″ apart, and use only a diametric band for testing. With a straightedge (see Fig. 55) and dividers, curvature amounting to 1/10 of the band separation, or 1/20 of a wave length of the light used, is easily measured. An appearance like that in Fig. 52, left, found on all three combinations, will indicate absolute flatness of all three disks. Zonal errors, disclosed by wiggles in the bands, are either depressions or ridges. the direction of the wiggle indicating which is the case. If caused by a ridge or hill, the curvature of the bands is convex toward the wedge opening between the disks; if caused by a depression or hollow, the bands are concave toward the wedge opening.

The precision of surface necessary depends on the purpose for which the flat is to be used. For the testing of diagonals, prism faces, filters, and sextant mirrors, flatness to a quarter wave (for a 6-inch disk) is acceptable. For a coelostat, or for the testing at the focus of achromatic objectives, paraboloidal mirrors, or Cassegrainian systems, the departure from flatness should not exceed 1/10 of a wave. One of the surfaces, or all three, should be figured to within these limits.

Appendix C

BIBLIOGRAPHY

OPTICS AND TELESCOPE MAKING

Amateur Astronomer's Handbook, J. B. Sidgwick. The Macmillan Company, New York, 1955.

Amateur Telescope Making — Book One, edited by Albert G. Ingalls. Scientific American Publishing Co., New York, 1955.

Amateur Telescope Making Advanced, edited by Albert G. Ingalls. Scientific American Publishing Co., New York, 1949.

Amateur Telescope Making — Book Three, edited by Albert G. Ingalls. Scientific American Publishing Co., New York, 1953.

Applied Optics and Optical Design, Alexander Eugen Conrady. Dover Publications, Inc., New York, Part 1, 1957; Part 2, 1960.

How To Make a Telescope, Jean Texereau. Interscience Publishers, Inc., New York, 1957.

Mirrors, Prisms and Lenses, James Powell Cocke Southall. The Macmillan Company, New York, 1933.

Telescopes: How To Make Them and Use Them, edited by Thornton Page and Lou Williams Page. The Macmillan Company, New York, 1966.

OBSERVING

American Ephemeris and Nautical Almanac. U. S. Government Printing Office, Washington, D. C. 20402, annual.

Field Guide to the Stars and Planets, Donald H. Menzel. Houghton Mifflin Company, Boston, 1964.

Handbook of the British Astronomical Association. British Astronomical Association, London, annual.

Handbook of the Constellations, Hans Vehrenberg and Dieter Blank. Sky Publishing Corporation, Cambridge, 1973.

Observational Astronomy for Amateurs, J. B. Sidgwick. The Macmillan Company, New York, 1955.

Observer's Handbook. Royal Astronomical Society of Canada, Toronto, Ontario, annual.

Skalnate Pleso *Atlas of the Heavens*, Antonin Becvar. Sky Publishing Corporation, Cambridge, 1962.

Star Atlas and Reference Handbook, A. P. Norton. Sky Publishing Corporation, 1973.

INDEX